Review of the Environmental Protection Agency's Draft IRIS Assessment of Tetrachloroethylene

Committee to Review EPA's Toxicological Assessment
of Tetrachloroethylene

Board on Environmental Studies and Toxicology

Division on Earth and Life Studies

NATIONAL RESEARCH COUNCIL
OF THE NATIONAL ACADEMIES

THE NATIONAL ACADEMIES PRESS
Washington, D.C.
www.nap.edu

THE NATIONAL ACADEMIES PRESS 500 Fifth Street, NW Washington, DC 20001

NOTICE: The project that is the subject of this report was approved by the Governing Board of the National Research Council, whose members are drawn from the councils of the National Academy of Sciences, the National Academy of Engineering, and the Institute of Medicine. The members of the committee responsible for the report were chosen for their special competences and with regard for appropriate balance.

This project was supported by Contract 68-C-03-081 between the National Academy of Sciences and the U.S. Environmental Protection Agency. Any opinions, findings, conclusions, or recommendations expressed in this publication are those of the authors and do not necessarily reflect the view of the organizations or agencies that provided support for this project.

International Standard Book Number-13: 978-0-309-15094-1
International Standard Book Number-10: 0-309-15094-9

Additional copies of this report are available from

The National Academies Press
500 Fifth Street, NW
Box 285
Washington, DC 20055

800-624-6242
202-334-3313 (in the Washington metropolitan area)
http://www.nap.edu

Printed in the United States of America.

THE NATIONAL ACADEMIES

Advisers to the Nation on Science, Engineering, and Medicine

COMMITTEE TO REVIEW EPA'S TOXICOLOGICAL ASSESSMENT OF TETRACHLOROETHYLENE

Members

SAM KACEW (*Chair*), University of Ottawa, Ontario, Canada
BRUCE H. ALEXANDER, University of Minnesota School of Public Health, Minneapolis
MARGIT L. BLEECKER, Center for Occupational and Environmental Neurology, Baltimore, MD
GARY P. CARLSON, Purdue University, West Lafayette, IN
LINDA D. COWAN, University of Oklahoma Health Sciences Center, Oklahoma City
MARY E. DAVIS, West Virginia University, Morgantown
H. CHRISTOPHER FREY, North Carolina State University, Raleigh
JOSEPH R. LANDOLPH, University of Southern California, Los Angeles
M.E. (BETTE) MEEK, University of Ottawa, Ontario, Canada
DAVID C. MCMILLAN, University of Nebraska Medical Center, Omaha
M. CHRISTOPHER NEWLAND, Auburn University, Auburn, AL
JULIA QUINT, California Department of Public Health (retired), Berkeley
GARY L. ROSNER, University of Texas M.D. Anderson Cancer Center, Houston
IVAN RUSYN, University of North Carolina, Chapel Hill
ROLF SCHULTE-HERMANN, Medical University of Vienna, Austria
IRVIN R. SCHULTZ, Battelle Pacific Northwest Division, Sequim, WA
ROBERT SNYDER, Rutgers, the State University of New Jersey, Piscataway
ROBERTA F. WHITE, Boston University School of Public Health, Boston, MA
LUOPING ZHANG, University of California, Berkeley
YILIANG ZHU, University of South Florida, Tampa

Staff

SUSAN N. J. MARTEL, Project Director
NORMAN GROSSBLATT, Senior Editor
MIRSADA KARALIC-LONCAREVIC, Manager, Technical Information Center
KEEGAN SAWYER, Associate Program Officer
TAMARA DAWSON, Program Associate
RADIAH ROSE, Editorial Project Manager

Sponsor

U.S. ENVIRONMENTAL PROTECTION AGENCY

OTHER REPORTS OF THE BOARD ON ENVIRONMENTAL STUDIES AND TOXICOLOGY

Contaminated Water Supplies at Camp Lejeune—Assessing Potential Health Effects (2009)
Review of the Federal Strategy for Nanotechnology-Related Environmental, Health, and Safety Research (2009)
Science and Decisions: Advancing Risk Assessment (2009)
Phthalates and Cumulative Risk Assessment: The Tasks Ahead (2008)
Estimating Mortality Risk Reduction and Economic Benefits from Controlling Ozone Air Pollution (2008)
Respiratory Diseases Research at NIOSH (2008)
Evaluating Research Efficiency in the U.S. Environmental Protection Agency (2008)
Hydrology, Ecology, and Fishes of the Klamath River Basin (2008)
Applications of Toxicogenomic Technologies to Predictive Toxicology and Risk Assessment (2007)
Models in Environmental Regulatory Decision Making (2007)
Toxicity Testing in the Twenty-first Century: A Vision and a Strategy (2007)
Sediment Dredging at Superfund Megasites: Assessing the Effectiveness (2007)
Environmental Impacts of Wind-Energy Projects (2007)
Scientific Review of the Proposed Risk Assessment Bulletin from the Office of Management and Budget (2007)
Assessing the Human Health Risks of Trichloroethylene: Key Scientific Issues (2006)
New Source Review for Stationary Sources of Air Pollution (2006)
Human Biomonitoring for Environmental Chemicals (2006)
Health Risks from Dioxin and Related Compounds: Evaluation of the EPA Reassessment (2006)
Fluoride in Drinking Water: A Scientific Review of EPA's Standards (2006)
State and Federal Standards for Mobile-Source Emissions (2006)
Superfund and Mining Megasites—Lessons from the Coeur d'Alene River Basin (2005)
Health Implications of Perchlorate Ingestion (2005)
Air Quality Management in the United States (2004)
Endangered and Threatened Species of the Platte River (2004)
Atlantic Salmon in Maine (2004)
Endangered and Threatened Fishes in the Klamath River Basin (2004)
Cumulative Environmental Effects of Alaska North Slope Oil and Gas Development (2003)
Estimating the Public Health Benefits of Proposed Air Pollution Regulations (2002)
Biosolids Applied to Land: Advancing Standards and Practices (2002)
The Airliner Cabin Environment and Health of Passengers and Crew (2002)

Arsenic in Drinking Water: 2001 Update (2001)
Evaluating Vehicle Emissions Inspection and Maintenance Programs (2001)
Compensating for Wetland Losses Under the Clean Water Act (2001)
A Risk-Management Strategy for PCB-Contaminated Sediments (2001)
Acute Exposure Guideline Levels for Selected Airborne Chemicals (seven volumes, 2000-2009)
Toxicological Effects of Methylmercury (2000)
Strengthening Science at the U.S. Environmental Protection Agency (2000)
Scientific Frontiers in Developmental Toxicology and Risk Assessment (2000)
Ecological Indicators for the Nation (2000)
Waste Incineration and Public Health (2000)
Hormonally Active Agents in the Environment (1999)
Research Priorities for Airborne Particulate Matter (four volumes, 1998-2004)
The National Research Council's Committee on Toxicology: The First 50 Years (1997)
Carcinogens and Anticarcinogens in the Human Diet (1996)
Upstream: Salmon and Society in the Pacific Northwest (1996)
Science and the Endangered Species Act (1995)
Wetlands: Characteristics and Boundaries (1995)
Biologic Markers (five volumes, 1989-1995)
Science and Judgment in Risk Assessment (1994)
Pesticides in the Diets of Infants and Children (1993)
Dolphins and the Tuna Industry (1992)
Science and the National Parks (1992)
Human Exposure Assessment for Airborne Pollutants (1991)
Rethinking the Ozone Problem in Urban and Regional Air Pollution (1991)
Decline of the Sea Turtles (1990)

Copies of these reports may be ordered from the National Academies Press
(800) 624-6242 or (202) 334-3313
www.nap.edu

Preface

In June 2008, the U.S. Environmental Protection Agency (EPA) released its draft *Toxicological Review of Tetrachloroethylene (Perchloroethylene) (CAS No. 127-18-4) in Support of Summary Information on the Integrated Risk Information System (IRIS)*. The assessment provided estimates of cancer and noncancer effects, which will be used to establish air and water quality standards to protect public health and set cleanup standards for hazardous-waste sites. EPA requested that the National Research Council review the scientific evidence on the adverse health effect of tetrachloroethylene and the agency's application of such data in quantifying human health risks. The review was sought to ensure that the draft IRIS assessment was consistent with current EPA guidance on conducting risk assessments and that it reflected sound scientific analysis and judgment.

In response to EPA's request, the National Research Council convened the Committee to Review EPA's Toxicological Assessment of Tetrachloroethylene, which prepared this report. The members of the committee were selected for their expertise in pharmacokinetics, liver toxicology, kidney toxicology, neurotoxicology, hematopoietic toxicology, reproductive toxicology, developmental toxicology, genotoxicity, carcinogenesis, epidemiology, physiologically based pharmacokinetic modeling, biostatistics, and risk assessment. Biographic information on the committee members is provided in Appendix A.

To help the committee in its review, public meetings were held in November 2008 and January and April 2009 to gather information from EPA, academic and industry researchers, state public-health departments, and the general public. The committee is grateful to those who gave presentations on research related to tetrachloroethylene or on topics relevant to the committee's task, including Judith Schreiber, Office of the New York State Attorney General; Philip Bushnell, EPA; Thomas Burke, Johns Hopkins Bloomberg School of Public Health; Andy Salmon, California Environmental Protection Agency; and Harvey Clewell III, Hamner Institutes for Health Sciences. The committee also thanks Peter Preuss, Kathryn Guyton, and Karen Hogan for providing background information and responding to questions throughout the study.

One committee member, Rolf Schulte-Hermann, disagreed with the committee's support of EPA's conclusion that the mode of action of tetrachloroethylene in inducing liver cancer in rodents is unknown. He judges that the induction of liver cancer in mice can be fully explained by a mode of action that involves the activation of the peroxisome proliferator-activated receptor-alpha. The basis of his judgment and of his dissent from the committee's position is detailed in Appendix B, where it is followed by the committee's rebuttal.

This report and the dissenting statement have been reviewed in draft form by persons chosen for their diverse perspectives and technical expertise in accordance with procedures approved by the National Research Council's Report Review Committee. The purpose of the independent review is to provide candid and critical comments that will assist the institution in making its published report as sound as possible and to ensure that the report meets institutional standards of objectivity, evidence, and responsiveness to the study charge. The review comments and draft manuscript remain confidential to protect the integrity of the deliberative process. We thank the following for their review of this report: A. John Bailer, Miami University; Lucio Costa, University of Washington; Scott E. Bowen, Wayne State University; Wolfgang Dekant, University of Würzburg; Adnan Elfarra, University of Wisconsin; Jeffrey Fisher, University of Georgia; David H. Garabrant, University of Michigan; Bernard D. Goldstein, University of Pittsburgh; David G. Hoel, Medical University of South Carolina; Ronald Melnick, National Institute of Environmental Health Sciences; Dorothy Patton, Environmental Protection Agency (retired); David Richardson, University of North Carolina School of Public Health; and Lauren Zeise, California Environmental Protection Agency.

Although the reviewers listed above provided many constructive comments and suggestions, they were not asked to endorse the conclusions or recommendations, nor did they see the final draft of the report before its release. The review of the report was overseen by the review coordinator, David Eaton, University of Washington, and review monitor, Mark Cullen, Yale University. Appointed by the National Research Council, they were responsible for making certain that an independent examination of the report was carried out in accordance with institutional procedures and that all review comments were carefully considered. Responsibility for the final content of the report rests entirely with the author committee and the institution.

The committee is grateful for the assistance of National Research Council staff in preparing the report, in particular Susan Martel, who served as project director and contributed to the report. Other staff members who contributed are James Reisa, director of the Board on Environmental Studies and Toxicology; Keegan Sawyer, associate program officer; Norman Grossblatt, senior editor; Mirsada Karalic-Loncarevic, manager of the Technical Information Center; Radiah Rose, editorial projects manager; and Tamara Dawson, program associate.

Finally, I thank all the members of the committee for their time and efforts throughout the development of this report.

Sam Kacew, *Chair*
Committee to Review EPA's
Toxicological Assessment of
Tetrachloroethylene

Abbreviations

AUC	area under the curve
BMC	benchmark concentration
BMCL	benchmark concentration with its lower confidence limit
BMD	benchmark dose
BuChE	butyrylcholinesterase
CCI	color-confusion index
CFU	colony-forming unit
CHO	Chinese hamster ovary
CI	confidence interval
CNS	central nervous system
CYP	cytochrome P-450
DCA	dichloroacetic acid
DEHP	diethylhexylphthalate
EBV	Epstein Barr virus
8-OHdG	8-hydroxydeoxyguanosine
EPA	U.S. Environmental Protection Agency
FDA	Food and Drug Administration
FMO	flavin-containing monooxygenase
GJIC	gap junctional intercellular communication
GSH	glutathione
GST	glutathione *S*-transferase
HD	Hodgkin disease
IARC	International Agency for Research on Cancer
IRIS	Integrated Risk Information System
JEM	job-exposure matrix
JISA	Japan Industrial Safety Association
JTEM	job-task exposure matrix
LGLL	large granular lymphocytic leukemia
LOAEL	lowest observed-adverse-effect level
MCL	mononuclear-cell leukemia
MOA	mode of action
***N*-Ac-TCVCS**	*N*-acetyl-*S*-(1,2,2-trichlorovinyl)-L-cysteine

NCI	National Cancer Institute
NES	Neurobehavioral Evaluation System
NHL	non-Hodgkin lymphoma
NK	natural-killer
NOAEL	no-observed-adverse-effect level
NRC	National Research Council
NTP	National Toxicology Program
OECD	Organisation for Economic Co-operation and Development
OR	odds ratio
PBPK	physiologically based pharmacokinetic modeling
PCO	palmitoyl-CoA oxidation
POD	point of departure
PPARα	peroxisome proliferator-activated receptor-alpha
RCC	renal-cell carcinoma
RfC	reference concentration
RfD	reference dose
RfV	reference value
SAB	Science Advisory Board
SCE	sister-chromatid exchange
SIR	standardized incidence ratio
SMR	standardized mortality ratio
TCA	trichloroacetic acid
TCVC	*S*-(1,2,2-trichlorovinyl)-L-cysteine
TCVCS	*S*-(1,2,2-trichlorovinyl)-L-cysteine sulfoxide
TCVG	*S*-(1,2,2-trichlorovinyl) glutathione
TWA	time-weighted average
VCS	visual-contrast sensitivity
VEP	visual evoked potential
WHO	World Health Organization

Contents

APPENDIXES

BOXES, FIGURES, AND TABLES

BOXES

FIGURES

TABLES

Review of the Environmental Protection Agency's Draft IRIS Assessment of Tetrachloroethylene

Summary

Tetrachloroethylene is a volatile, chlorinated organic hydrocarbon that is widely used as a solvent in the dry-cleaning and textile-processing industries and as an agent for degreasing metal parts. It is an environmental contaminant that has been detected in the air, groundwater, surface waters, and soil. In June 2008, the U.S. Environmental Protection Agency (EPA) released its draft *Toxicological Review of Tetrachloroethylene (Perchloroethylene) (CAS No. 127-18-4) in Support of Summary Information on the Integrated Risk Information System (IRIS)*. The draft IRIS assessment provides quantitative estimates of cancer and noncancer effects of exposure to tetrachloroethylene, which will be used to establish air-quality and water-quality standards to protect public health and to set cleanup standards for hazardous-waste sites.

At the request of EPA, the National Research Council convened a committee to conduct an independent scientific review of the draft IRIS assessment of tetrachloroethylene from toxicologic, epidemiologic, and human clinical perspectives. The committee was asked to evaluate the adequacy of the EPA assessment, the data and methods used for deriving the noncancer values for inhalation and oral exposures and the oral and inhalation cancer unit risks posed by tetrachloroethylene; to evaluate whether the key studies underlying the draft IRIS assessment are of requisite quality, reliability, and relevance to support the derivation of the reference values and cancer risks; to evaluate whether the uncertainties in EPA's risk assessment were adequately described and, where possible, quantified; and to identify research that could reduce the uncertainty in the current understanding of human health effects associated with tetrachloroethylene exposure.

COMMITTEE'S ASSESSMENT

The committee appreciates the extensive work that EPA has invested in the development of its draft assessment of tetrachloroethylene. However, the committee has identified concerns about some of the approaches that EPA used to evaluate the data on tetrachloroethylene and subjects about which inadequate information or rationales are used to support its risk assessment—factors that call into question the soundness and reliability of EPA's proposed reference values and cancer risk estimates for tetrachloroethylene. One of the overarching weaknesses of the draft assessment was a lack of critical analysis of the data on which EPA relied in evaluating methodologic strengths and weaknesses. That lack was particularly evident in the assessment of the epidemiologic data: study selection and conclusions appeared to be based heavily on results that showed positive associations, and other data and the strengths and weaknesses of the selected studies were not adequately taken into consideration. The committee observed similar problems in its review of EPA's evaluation of the genotoxicity evidence, in which preference appeared to be given to studies that reported positive results. Specifically, EPA did not analyze studies critically with respect to their methodologic strengths and weaknesses, nor did it organize its discussion clearly to provide an integrated consideration of the weight of evidence on the genotoxicity of tetrachloroethylene. Other mode of action evaluations were also hampered in this way.

In the sections below, the committee evaluates EPA's noncancer and cancer assessments of tetrachloroethylene. The committee's recommendations focus on improvements that should be made by EPA in producing its final assessment and on improvements that EPA should pursue in the future when tetrachloroethylene is due for another update.

Noncancer Assessment

For noncancer effects of tetrachloroethylene, EPA proposes to set an inhalation reference concentration (RfC) and oral reference dose (RfD). Those are estimates (with uncertainty spanning perhaps an order of magnitude) of a continuous inhalation exposure and a daily oral exposure of the human population (including sensitive subgroups), respectively, that are likely to be without appreciable risk of deleterious effects during a lifetime. EPA's proposed RfC is 0.016 mg/m^3 (2 ppb), and its proposed RfD is 0.004 mg/kg per day. Those values are based on the neurobehavioral outcomes of visual dysfunction and cognitive deficits observed in epidemiologic studies. A 1995 study by Altmann et al., in which adverse neurotoxic effects (as measured by deficits in vigilance, reaction time, and visual memory) were observed in people who lived near dry-cleaning facilities, was selected as the basis of the derivation of the RfC and RfD. The committee was asked to evaluate the selection of neurobehavioral outcomes in support of the RfC and RfD, the key study used, approaches to route-to-route extrapolation, and the characterization of the uncertainties associated with the data.

Critical Noncancer End Point and Studies

The committee found that EPA adequately supported its selection of neurotoxicity as the critical effect on which to base the RfC and RfD. The draft IRIS document illustrates that neurotoxic effects are the most sensitive effects of tetrachloroethylene and that reference values based on neurotoxic effects would be protective against other noncancer effects that occur at higher concentrations.

EPA provides descriptions of the relevant neurotoxicity studies, but its evaluation of the epidemiologic literature could be improved by providing a critical evaluation of the validity of study designs and evaluation of the methods used for data collection and analysis, which the committee judges to be most important in selecting key studies. EPA chose the 1995 study by Altmann et al. as the critical one for determining the RfC and RfD because it involved an environmental exposure and used a standardized computer-assisted testing battery. Those are reasonable bases for the choice, but they do not outweigh methodologic deficiencies that seriously compromised the results of the study. Most important, the referent group was not appropriate. The group had more education than the exposed group and appeared to have pre-existing differences in cognitive abilities, which could account for its better test results. Evidence of residual confounding by education can be seen in the variability in reported results. For example, there was no association between tetrachloroethylene and visual evoked potentials; this is important because changes in the visual system and abnormalities in visual evoked potentials have been associated with tetrachloroethylene and other related solvents, and they are essentially unrelated to education. Other limitations of the study included the lack of a rationale for initial selection of study subjects, inadequacy of exposure characterization, and lack of a dose-response relationship. Finally, even though the test battery was performed properly, some of the tests have not been well validated with regard to what they reveal about brain damage.

Thus, the committee disagrees with EPA's selection of the 1995 Altmann et al. study as the basis of its risk calculations. In reviewing the database, the committee gave greater weight to studies that had the strongest methods; it neither chose nor excluded studies on the basis of their results. The set of studies that the committee judged to be more appropriate for supporting the RfC and RfD include those of Altmann et al. (1990), Cavalleri et al. (1994), Gobba et al. (1998), Echeverria et al. (1995), and Boyes et al. (2009).

Derivation of Reference Values

EPA derived sample inhalation reference values by using results from several supporting neurotoxicity studies for comparison with its principal study by Altmann et al. The committee found that some uncertainty factors were applied inconsistently; specifically, the application of the uncertainty factor to account for subchronic exposures in epidemiologic studies should be justified better. In some cases, EPA did not use such a factor; in other cases, it applied a value of 10 with weak justification.

The committee derived candidate values by using the same studies as EPA and additional studies. The committee found that the reference values from the strongest studies were in the range of 6-50 ppb (or 0.04-0.34 mg/m^3). That range is higher than the RfC of 0.016 mg/m^3 derived by EPA and is further supported when considered in the context of the full database (see further discussion below).

EPA extrapolated the results of inhalation studies to derive the oral RfD for tetrachloroethylene. Physiologically based pharmacokinetic (PBPK) modeling was used to support the route-to-route extrapolation. The rationale behind that approach is sound and adequately explained by EPA, and the choice of dose metric (blood area-under-the-curve) was appropriate and adequately supported by the available evidence. However, the three models used by EPA were formulated and validated with data from inhalation exposures; none was validated against blood concentrations that result from oral exposure. EPA empirically assumed a value for the rate of oral absorption of tetrachloroethylene; this assumption is inferior to direct estimation. Other PBPK models that use direct estimation are available, and their use may help to reduce the uncertainty in the assumed values; or additional PBPK models could be developed (see recommendation below for a harmonized PBPK model).

Graphical Presentation

EPA provides graphical comparisons of reference values, values that could be derived from supporting studies. Reference values derived from neurotoxicity data are presented, as are values based on other noncancer effects to illustrate dose dependence of multiple forms of observed toxicity. Overall, the committee supports the approach of presenting the evidence in this visual format. However, the committee recommends some revisions to improve illustration of the uncertainties being represented and to expand the presentation to include the larger body of literature on a particular end point to show how the RfC compares with sample reference values derived from studies that are methodologically sound but not judged to be critical for the RfC. Consistency between the RfC and such studies would provide additional support.

Figure S-1 provides an example illustration developed by the committee. It shows that the majority of sample values is centrally clustered, but there is a wide spread at the lower and higher ends. The overall range of the 19 sample reference values is 0.03-333 ppb (0.0002-2.6 mg/m^3), but the range is reduced to about 6-50 ppb (0.04-0.34 mg/m^3) when consideration is restricted to the five strongest studies. The RfC of 0.016 mg/m^3 calculated by EPA on the basis of the 1995 Altmann et al. study falls below the range. The figure shows that sample reference values that could be derived from the full database of neurotoxicity studies provide some support for the range.

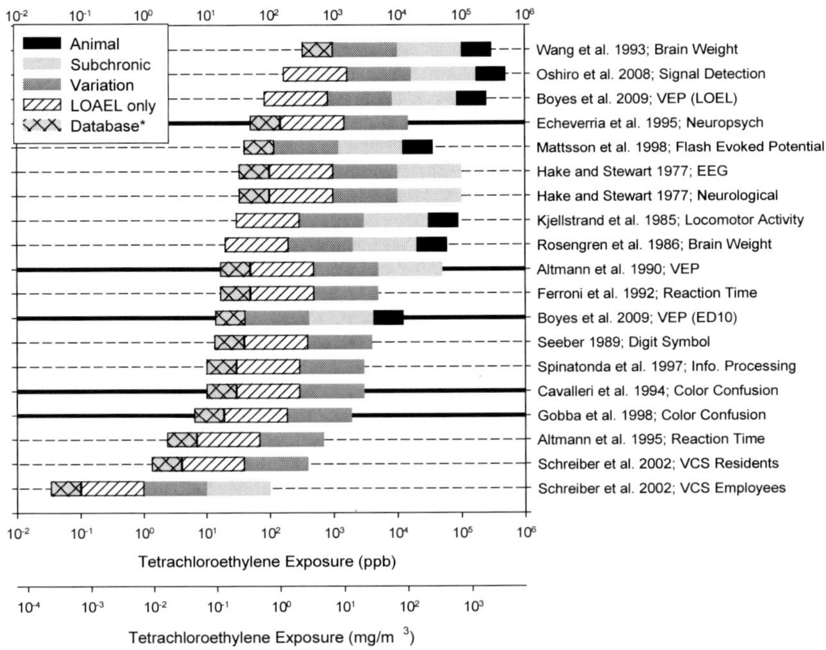

FIGURE S-1 Distribution of sample reference values. Each horizontal bar represents a single study. Thick, horizontal lines represent studies identified by the committee as most applicable to the development of an RfC. The right end of a bar is at the "point of departure" and is based on concentrations used in the referenced study after conversion to "human equivalencies" or, in the case of animal studies, after adjustment for continuous exposure. Uncertainty factors are illustrated in different shadings: a factor of 3 if it is necessary to extrapolate from animals to humans (black); a factor of 10 if it is necessary to extrapolate from acute or subchronic exposure to chronic exposure (light gray); a factor of 10 for individual variation to account for sensitive individuals (dark gray); a factor of 10 if the study did not contain a NOAEL (diagonal lines) and a factor of 3 for uncertainty in the data base as applied by EPA (light gray, cross-hatched). *A maximum total uncertainty factor of 3,000 was applied for the purpose of this exercise. Where this might be exceeded, the maximum was achieved by omitting the "database" uncertainty so that other uncertainties could be visualized. The committee has recommended that EPA review the uncertainty factors to ensure that they are appropriately explained and used consistently, so some of the individual values used here could be subject to change. In some cases, EPA might judge that the total uncertainty exceeds 3,000 and would, therefore, not use that study to derive a sample reference value. Source: Graphic developed by M. Christopher Newland.

Cancer Assessment

EPA faced a formidable challenge in its effort to characterize the carcinogenic properties of tetrachloroethylene both qualitatively and quantitatively.

There appears to be general agreement in the scientific community that tetrachloroethylene is carcinogenic in laboratory animals, but there is a longstanding debate about how to interpret and use the laboratory findings to predict human cancer risks. The debate is reflected in the committee's inability to reach consensus on some aspects of the tetrachloroethylene assessment, which are discussed below.

Classification

EPA classified tetrachloroethylene as "likely to be carcinogenic to humans." The committee reviewed the classification guidance in EPA's 2005 *Guidelines for Carcinogen Risk Assessment* and the bioassay data available on tetrachloroethylene and concluded that EPA adequately documented that its classification has been based on the results of bioassays that found increased incidences of hepatocelluar tumors, mononuclear-cell leukemia (MCL), renal tumors, and hemangiosarcomas in laboratory animals and to a lesser extent on epidemiologic evidence. EPA's decision to characterize tetrachloroethylene as likely to be a human carcinogen as opposed to "carcinogenic to humans" appropriately reflects the possibility that there are deficiencies or potential inaccuracies in interpretation of the data. Some of the possible deficiencies and inaccuracies are discussed below for each of the datasets.

Mononuclear-Cell Leukemia

An increased incidence of MCL in F344 rats has been reported in two bioassays. The biologic significance of the increases was debated by the committee because increases were observed in only one strain of rat, which is known to have a high background incidence of MCL, and because MCL's relevance to humans and the mode of action of tetrachloroethylene causing it are not understood. In considering the high background of MCL, the committee found a published assessment by Thomas et al. (2007) that applied statistical approaches (life-table analyses) to bioassays of the National Toxicology Program (NTP) to interpret dose response relationships. Tetrachloroethylene was one of five chemicals of 500 tested by NTP that showed statistically significant increases in MCL in both male and female rats despite the high background rates. The publication advocated that such statistical evidence be supported with a weight-of-evidence analysis of biologic data before conclusions were drawn.

The committee found some support from epidemiologic studies that suggested an association between tetrachloroethylene and lymphoma, but the data were relatively weak and inconsistent. A difficulty in interpreting the findings is a difference of opinion about the human relevance of MCL. Some committee members judged that similarities between a form of human leukemia (natural killer-cell large granular lymphocyte leukemia) and rat MCL and results of mechanistic studies that the committee recommended be added to EPA's assessment were adequate to establish human relevance; others believed that more research was needed to establish the relevance. The committee agreed that there was little information on a mode of action of tetrachloroethylene in increasing MCL and that it therefore was not possible to determine whether exposure to tetrachloroethylene results in initiation of new tumors or enhances the expansion or promotion of existing tumors.

Hepatic Cancer

Statistically significant increases in hepatic tumors were observed in male and female mice after oral or inhalation exposure. As in the case of MCL, the biologic significance of the increases was debated by the committee because $B6C3F_1$ mice have a high background incidence of hepatic cancer. However, the findings were reproduced in several studies conducted in different laboratories and showed a dose-response relationship. There is also fairly substantial information for characterizing potential modes of action of hepatic-tumor formation relative to the data available on MCL and renal cancer. Although the committee recommended that EPA revise its presentation of the mode-of-action evidence on tetrachloroethylene-related hepatic cancer to clarify its position, most of the members agreed with EPA that the mode of action is complex and remains to be established. The latter members also agreed that there was insufficient evidence to rule out human relevance. One member objected to those conclusions and to the committee's support of using hepatic cancer to quantify risk. He argued that in the absence of evidence of other contributing modes of action, the evidence is sufficient to conclude that the mode of action in mice is predominantly through activation of the peroxisome proliferator-activated receptor-alpha, a mode of action that he considered to be of little relevance to humans. His arguments are presented in a dissenting statement in Appendix B of the report.

Renal Cancer

Tetrachloroethylene caused a low rate of induction of renal tumors in rats. Although the increases were not statistically significant when compared with concurrent controls, EPA has used historical controls to calculate the chances of two of these rare carcinomas to occur by chance to be less than 0.001. Furthermore, a dose-response trend was shown against the low background and the tumors in the treated rats were malignant whereas the tumors in the controls were not. EPA provided a strong evaluation of the potential modes of action for tetrachloroethylene-induced kidney cancer. The committee agrees with EPA that the mode of action of tetrachloroethylene tumorigenesis is not understood but that a mutagenic mode of action cannot be ruled out. Thus, renal tumors observed in tetrachloroethylene-treated rats were considered relevant to humans although additional characterization of quantitative relevance is desirable.

Selection of Tumor Type for Quantitative Assessment

The committee was unable to reach consensus on the selection of the critical cancer end point. The majority of the members judged that the uncertainties associated with MCL (particularly the high background incidence, uncertainty about the dose-response relationship, and poor understanding of mode of action) were too great to support using MCL data rather than data on hepatic or renal cancer for determining quantitative estimates of risk. Those members judged that the use of the MCL data could be justified only if it is EPA's policy to choose the most conservative unit risk when considering options but that such justification should be distinguished as a policy decision, not a scientific one. They believed that a more scientifically defensible approach would be to use the dataset that has the least uncertainty rather than the dataset that yields the highest estimate of risk. In their judgment, the hepatic-cancer data would have the least uncertainty, followed by the data on renal cancer and MCL.

Other members judged that the MCL data should be used for cancer-risk estimation. Their opinions were based on the observation that reproducible, statistically significant increases in MCL in male and female rats above the background incidence of MCL were found and that MCL was the cancer end point with the highest magnitude of response. They believed that use of the most sensitive response to quantify cancer risk decreases the uncertainty associated with potential differences in metabolism and susceptibility to tetrachloroethylene among exposed populations. They concluded that additional statistical analyses of the dose-response data and the addition of supporting mechanistic information identified by the committee would strengthen the existing support of the use of MCL in the draft assessment.

Mode-of-Action Considerations

The modes of action[1] by which tetrachloroethylene produces increases in

---

[1]There was some disagreement among the committee members on what constitutes "modes of action" and "key events." In Section 4.4.4 of the draft IRIS assessment, EPA discusses several "topics" relevant to the mode of action for hepatic toxicity, including metabolism, receptor activation, genotoxic effects, and nongenotoxic effects. EPA's presentation treats those topics as separate modes of action, but metabolism is presented as a key event or a component of multiple modes of action. Some committee members judged that that treatment was appropriate as an introduction to a discussion of multiple modes of action and was consistent with EPA guidelines. Other members judged that although early key events may occur in different pathways, they converge to produce one effect; thus, these members hold the view that there is one mode of action for an observed effect for which there are a number of specified key events (early key events may be derived from a series of pathways). Despite those differing viewpoints, all members of the committee agreed that more focused analyses of the available evidence are necessary to support hypothesized modes of action.

MCL, hepatic cancer, and renal cancer were an important consideration in EPA's and the committee's evaluations of the evidence. The analytic framework described in EPA's cancer guidelines for considering hypothesized modes of action was best applied in the draft IRIS assessment's consideration of renal cancer. The evaluation focused on synthesizing the evidence to support the idea that multiple modes of action may play a role. However, for hepatic cancer, the committee found that the assessment lacked the organization to present and provide appropriate context for the evidence clearly. It therefore recommended that EPA revise its mode-of-action assessment for hepatic cancer to support better the conclusions that were drawn. Specifically, the committee suggested that the mode-of-action analyses would be improved by outlining the proposed sequence of hypothesized tetrachloroethylene-associated key events (possibly with a diagram). Transparency would be improved by presenting the details of experimental results in tabular form to allow the reader to understand more easily the relative potency of tetrachloroethylene, or its metabolites, in inducing both key events and tumors. In this context, species and strain differences could also be considered more easily. The goals of the presentation should be to lay out the timeline of key events explicitly in the context of dose, to evaluate concordance between early and late events, and to consider the relative contribution of chemical-specific data compared with information on categories of chemicals. This approach should be applied to each hypothesized mode of action. Even if the data are ultimately judged to be insufficient to support a hypothesis, the exercise can be used to identify critical data gaps and to inform the direction of future research.

Low-Dose Extrapolation

EPA's dose-response analyses of the various cancer datasets involved using several models to extrapolate to doses below the experimental range. EPA considered six datasets: hepatocellular adenoma or carcinoma in male and female mice, hemangiosarcoma in male mice, MCL in male and female rats, and renal tumors in male rats. It used the multistage model for each dataset because mode-of-action information was lacking or uncertain and the model was able to fit a broad array of dose-response patterns. However, because the studies used small numbers of dose groups and because the benchmark-dose software automatically fixed some parameters to zero to obtain convergence in model-fitting, the fitted models were nearly linear in the low-dose range. The imposed linearity explains the similarity among the slopes of the models and among the unit risks derived from the models. In the case of hepatocellular adenoma and carcinoma in male mice and MCL in female rats, EPA considered the fitted models acceptable solely on the grounds that statistical tests for goodness of fit had nonsignificant results ($p > 0.10$). The committee considers this to be a weak rationale in that the statistical significance of goodness-of-fit tests may not detect a poor fit when the number of animals per dose group is small. The questionable fitting of the multistage model to some candidate datasets and insufficient consideration of alternative models contribute to underestimation of the overall uncertainties.

EPA adopted linear low-dose extrapolation, the default option, with several justifications. First, nonlinear, mechanistic models are unavailable for dose-response modeling because mode-of-action information on tetrachloroethylene is insufficient and support for dynamic models is unavailable. Second, because mathematical models are subject to uncertainties for low-dose extrapolation beyond the experimental dose range, linear extrapolation is more conservative than all sublinear (curvilinear) models. When individual thresholds in the human population are plausible, wide variation in threshold values typically implies a curvilinear shape of the dose-response relationship. Thus, linear extrapolation protects susceptible subpopulations. Third, a few of the candidate data, especially EPA's preferred male-rat MCL data, exhibit a linear dose-response relationship. Whereas those arguments are consistent with EPA's *Guidelines for Carcinogen Risk Assessment*, there is evidence in the candidate datasets that the underlying dose-response relationship can be supralinear (for example, in MCL in female rats). When that is the case, low-dose linear extrapolation is not conservative. EPA does not present the full ranges of variation and uncertainty in relation to model choice, in large part because it applied only linear or nearly linear dose-response models to all candidate datasets.

Age-Adjustment Factor

EPA did not apply an age-adjustment factor to its cancer risk assessment, because there is little evidence that tetrachloroethylene or its oxidative metabolites directly damage DNA, because information about genotoxicity of glutathione (GSH) metabolites in cell assays other than *Salmonella* or in vitro experiments is lacking, and because the mode of action of tetrachloroethylene has not been established. In addition, there are no data on differential sensitivity to tetrachloroethylene carcinogenicity among life stages. The committee agrees that those are adequate reasons for not using an age-adjustment factor but suggests that the rationale can be strengthened if EPA follows the committee's suggestions for improving its analysis of the genotoxicity data and mode-of-action evidence.

Physiologically Based Pharmacokinetic Models

Tetrachloroethylene can be viewed as being metabolized by three pathways. The predominant pathway is the cytochrome P-450 (CYP) pathway that produces metabolites that have been associated with hepatic cancer. Two other pathways involve the GSH conjugation pathway that produces metabolites that are further metabolized by the β-lyase pathway or the β-lyase-independent pathway, each of which produce metabolites that have been associated with renal cancer. To take those metabolic factors into account, EPA used three PBPK models to estimate human equivalent doses from animal studies and to perform route-to-route extrapolations. Each of the models used total metabolism of tetrachloroethylene as the dose metric. In some instances, EPA used a single model; in others, it used all three. The justification for using single or multiple models is not always clear. The committee observed that the models could yield different results because they were calibrated with different datasets, so comparisons among them were not straightforward. For consistency and to allow for better comparisons among end points, the committee recommends that EPA use a single PBPK model for its assessment. Ideally, the model would be a "harmonized" version of the three models used by EPA or of other relevant models (that is, a single model that integrates multiple exposure routes and tissue compartments).

The committee notes that the use of total metabolism as the dose metric for carcinogenicity reflects primarily the CYP metabolic pathway because of large differences in the flux of the metabolism between it and the GSH pathway. Using that dose metric does not reflect the contribution of the GSH conjugation pathway, which has been implicated in the development of renal cancer. EPA did not pursue the addition of the GSH pathway to any of the PBPK models, arguing that data on GSH-dependent metabolism are from in vitro studies or constitute measurements of urinary excretion products and do not represent toxic species in vivo. The committee agrees that the available data on the GSH pathway are more limited than the available data on the CYP pathway but notes that in vitro and urinary metabolite data were used in the development of the CYP-based PBPK models chosen by EPA. Thus, better justification is necessary to rule out modeling the GSH pathway.

The committee recommends that EPA explore the possibility of adding the GSH pathway to a harmonized PBPK model. If such modeling is determined to be infeasible, total metabolism can be used as a reasonably conservative dose metric. The modeling exercise would be useful in identifying data gaps that prevent successful modeling, which can be used to guide research that will allow more comprehensive PBPK models to be developed in support of the next IRIS reassessment of tetrachloroethylene.

Uncertainty Analysis

EPA has clearly identified key sources of uncertainty as part of its process of assessing the cancer risk posed by exposure to tetrachloroethylene, including human population variation, low-dose extrapolation, dose metrics, extrapolation from animals to humans, and the use of PBPK models for route-to-route extrapolation. The effect of uncertainties on risk estimates is assessed qualitatively in most parts of the IRIS draft except in dealing with such issues as the choice of dose-response models, the use of PBPK models, and, to a small degree, variation between studies. That approach reflects the current state of practice of uncertainty analysis.

In a few respects, the committee disagrees with EPA's presentation on uncertainties. For example, EPA notes narrow variation between cancer risks derived from four dose-response models. However, in its comparison, EPA used only data on male rats, and all four models were linear or nearly linear at lower doses. Failure to consider a wider array of feasible dose-response models, including multistage models of various orders, could lead to inadequate quantification of uncertainty associated with the choice of dose-response model.

The committee supports EPA's quantitative assessments of uncertainty with regard to choice of dose-response models, the use of PBPK models, and variation between studies. In particular, the committee found EPA's consideration of uncertainty due to different forms of dose-response models to be valuable, and it recommends that such quantitative evaluations be extended to all candidate datasets so that a fuller array of uncertainties can be assessed.

CONSIDERATIONS FOR FUTURE RE-EVALUATIONS OF TETRACHLOROETHYLENE

The committee found several parts of the draft IRIS assessment that could be improved on in the future. Such changes are not necessary for completing the current assessment but should be considered when tetrachloroethylene is reevaluated in the future. They include improving transparency in selection and analysis of data, particularly with regard to uncertainty analysis. The committee encourages EPA to consider the most recent guidance from the National Research Council report *Science and Decisions*.

Organization and Approach

There is a vast amount of literature on tetrachloroethylene, and the draft IRIS assessment was hampered by having to manage the sheer volume of information on the chemical. Any new reassessment should begin with problem formulation and issue identification, consideration of whether to rely on previous reviews, determination of the focus of the new effort, and identification of the specific issues on which the reassessment is likely to focus. That would help to identify where multidisciplinary input at early stages of reanalysis should be sought, such as in data selection and mode-of-action evaluations in the context of risk-assessment practices. The process would include a delineation of criteria for selecting studies, approaches for conducting a weight-of-evidence evaluation, and options for dose-response assessment and the characterization of uncertainties. EPA should also consider ways to reorganize the document to streamline presentation of the data and analyses.

Uncertainty Analysis

EPA's assessment of tetrachloroethylene follows a traditional approach to developing cancer slope factors and hazard indexes that takes uncertainties into account qualitatively and via uncertainty factors. EPA states that it has introduced a new method for uncertainty analysis in the context of the dose-response assessments for tetrachloroethylene, but the only notable differences between its tetrachloroethylene assessment and those of other chemicals are the consideration of multiple end points and the limited use of bootstrap simulation for only a portion of uncertainties. EPA's uncertainty analysis remained typically focused on individual sources of uncertainty, and the analysis was often qualitative without presenting a full range of the uncertainty. Without an in-depth illustration of the propagation and cumulative effect of the uncertainties on the final risk estimate, quantification of the overarching uncertainty surrounding the final risk assessment is not possible. The committee notes that the current state of practice in quantitative uncertainty analysis does not fully meet the spirit of principles, guidelines, and recommendations that have accrued in recent years.

1

Introduction

Tetrachloroethylene is a volatile chlorinated organic hydrocarbon that is widely used as a solvent in the dry-cleaning and textile-processing industries and as an agent for degreasing metal parts. It is also used as a chemical precursor for synthesis of fluorocarbons. It has the following use pattern: 55% as a chemical intermediate, 25% for metal-cleaning and degreasing, 15% for dry-cleaning and textile-processing, and 5% for other unspecified uses (ATSDR 1997; EPA 2008). Dry-cleaning facilities are an important source of atmospheric emissions of tetrachloroethylene. Tetrachloroethylene becomes a groundwater contaminant as a result of leaks and improper disposal practices; it can persist in groundwater for years because it has little contact with air. The U.S. Environmental Protection Agency (EPA) has classified tetrachloroethylene as a hazardous air pollutant under the Clean Air Act, a toxic pollutant under the Clean Water Act, a contaminant under the Safe Drinking Water Act, a hazardous waste under the Resource Conservation and Recovery Act, and a hazardous substance under the Comprehensive Environmental Response, Compensation, and Liability Act.

EPA's Integrated Risk Information System (IRIS) is a database that provides the agency's assessments of potential human health effects of exposure to various substances in the environment. IRIS assessments provide quantitative estimates of cancer and noncancer effects that are used to establish air and water quality standards to protect public health and set cleanup standards for hazardous-waste sites. For noncancer effects, EPA establishes an oral reference dose (RfD) and an inhalation reference concentration (RfC), which are estimates (with uncertainty spanning perhaps an order of magnitude) of daily oral exposure and continuous inhalation exposure of the human population (including sensitive subgroups), respectively, that are likely to be without an appreciable risk of deleterious effects during a lifetime. For cancer, the IRIS database provides a characterization of the weight of evidence of human carcinogenicity, oral slope factors, and inhalation unit risks. An oral slope factor is an upper bound, approximating a 95% confidence limit, on the increased cancer risk posed by lifetime exposure to an agent; it is usually expressed in units of proportion (of a population) affected per milligram per kilogram of body weight per day. A unit risk is the upper bound on the excess lifetime cancer risk estimated to result from continuous exposure to an agent at a concentration of 1 µg/L in water or 1 µg/m^3 in air. For example, a unit risk of 2×10^{-6} per microgram per liter is interpreted as 2 excess cancer cases (upper-bound estimate) expected to develop per 1,000,000 people if they are exposed to the chemical daily for a lifetime at 1 µg per liter of drinking water.

EPA requested that the National Research Council undertake an independent assessment of its draft *Toxicological Review of Tetrachloroethylene (Perchloroethylene) (CAS No. 127-18-4) in Support of Summary Information on the Integrated Risk Information System (IRIS),* hereafter called the draft IRIS assessment. The draft IRIS assessment proposes an RfC of 1.6×10^{-2} mg/m^3, an RfD of 4×10^{-3} mg/kg-day, a range of inhalation unit risks of 2×10^{-6} to 2×10^{-2} per mg/m^3, and a range of oral slope factors of 1×10^{-2} to 1×10^{-1} per mg/kg-day. EPA requested a review of those values and their scientific basis in 2006 but delayed public release of the draft IRIS assessment for additional evaluation within the agency. Therefore, the committee's review did not begin until June 2008, when the draft was released.

STATEMENT OF TASK

A committee convened by the National Research Council was asked to conduct a scientific review—from toxicologic, epidemiologic, and human clinical perspectives—of EPA's draft IRIS assessment of tetrachloroethylene that was made available for external review. The committee's review was to include an evaluation of the adequacy of the assessment and the data and methods used for deriving the RfD and RfC of tetrachloroethylene and its oral and inhalation cancer unit risks. The committee was asked to evaluate whether the key studies underlying the draft IRIS assessment were of requisite quality, reliability, and relevance to support the derivation of the RfD, RfC, and oral and inhalation unit risks; to evaluate whether the scientific uncertainties in EPA's risk assessment were adequately described and, where possible, quantified; and to identify research that could reduce the uncertainties given the current understanding of human health effects associated with tetrachloroethylene exposure.

During the study course of the project, EPA submitted specific questions for the committee to address. The final list, submitted in February 2009, included the following questions:

General Charge Questions:

1. Does the draft IRIS assessment provide a scientifically sound, balanced, and transparent review and synthesis of the key scientific evidence on chronic noncancer and cancer hazard and risk?

2. Please identify any additional important studies that should be considered in the assessment of the chronic noncancer and cancer health effects of tetrachloroethylene.

Specific Charge Questions:

Noncancer Assessment

1. Selection of neurotoxicity as the basis for the RfC and RfD for tetrachloroethylene—a number of studies assessing neurobehavioral and other effects in both humans and rodents are available for RfC and RfD analysis.

   a. Is EPA's selection of neurotoxicity, specifically visual dysfunction and cognitive deficits, appropriate for providing a point of departure for derivation of the RfC and RfD? The goal of a reference value is to provide an estimate of exposure of the human population (including susceptible subgroups) that is likely to be without appreciable risk of adverse health effects over a lifetime.
   b. Does EPA provide a sound and transparent description of the relevant studies of the neurotoxic effects of tetrachloroethylene?
   c. Does the assessment present an appropriate rationale for selection of the study by Altmann et al. (1995) as the critical study? If another study is judged more appropriate for use as the critical study, please provide a critical evaluation of it and of its suitability for meeting the goals of a reference value.

2. Characterization of Uncertainties—the noncancer assessment considers uncertainty on the basis of extrapolation from laboratory animals to humans, variations in response within experimental species, human variation, and database deficiencies; the noncancer RfC and RfD are based on a specific neurotoxicity effect; EPA also presents reference values based on other effects to illustrate the dose dependence of the multiple observed toxicities.

   a. Has EPA accurately and clearly characterized the basis of selection of uncertainty factors for the RfC and RfD? Please comment on the rationales underlying the choice of uncertainty factors, such as the database uncertainty factor, which is intended to account for the degree of limitations in both human and animal data.
   b. Please comment on EPA's graphic presentation of noncancer reference values that could have been derived from studies of different neurotoxic effects or toxic effects in other organ systems.

Cancer Assessment

1. Weight-of-evidence descriptor—the assessment concludes that tetrachloroethylene is "likely to be carcinogenic to humans" by all routes of exposure

within the framework of the *Guidelines for Carcinogen Risk Assessment* (EPA 2005a).

a. Does EPA provide a clear and cogent weight-of-evidence evaluation?
b. Does the assessment support the conclusion that tetrachloroethylene by oral and inhalation exposure is likely to be carcinogenic in humans (at all levels of exposure)?

2. Mode of action considerations—the mode of action of a carcinogen can inform identification of hazards and approaches used for a dose-response relationship; the assessment concludes that a mode of action of tetrachloroethylene has not been definitively established for any of the site-specific tumor types.

a. Does EPA provide a sound evaluation and characterization of the available data related to mode(s) of action for the carcinogenicity of tetrachloroethylene?
b. Do the available data support EPA's conclusion that mode(s) of action for tetrachloroethylene-induced carcinogenesis is unknown?
c. Does EPA clearly address why age-dependent adjustment factors for cancer risk are not applied, according to the *Guidelines for Carcinogen Risk Assessment* (EPA 2005a) and *Supplemental Guidance for Assessing Cancer Susceptibility from Early-Life Exposure to Carcinogens* (EPA 2005b)?

3. Development of the inhalation unit risk and oral slope factor—EPA's draft unit-risk estimate relies on choices of tumor type, point of departure, and low-dose extrapolation that aim to provide a "reasonable upper bound estimate" of risk; because the draft assessment judged that there was no strong basis for preferring one physiologically-based pharmacokinetic model over another, a range of tetrachloroethylene unit-risk estimates calculated with three PBPK models is given.

a. Please comment on EPA's selection of mononuclear-cell leukemia in male rats from the Japanese Industrial Safety Association study for quantitative derivation of the inhalation unit risk and oral slope factor. Note that, consistently with the *Guidelines for Carcinogen Risk Assessment* (EPA 2005a), the draft IRIS assessment does not infer site concordance of tumors across species. If another study or end point is judged to be more appropriate for the derivation of these risk values, please provide a critical evaluation of the end point and its suitability for supporting a unit risk estimate.
b. Does EPA clearly and objectively describe the low-dose extrapolation approach, that is, linear extrapolation in accordance with default recommendations in the *Guidelines for Carcinogen Risk Assessment* (EPA 2005a)?

4. Consideration of uncertainties—the cancer assessment considered the contribution of a number of sources of uncertainty; some uncertainties (for example, pertaining to mode of action and human sensitivity and variability) were qualitatively expressed, and in other cases EPA examined the potential quantita-

tive impact on the risk estimate; in addition to the unit risk estimate, the assessment provides lower bounds (such as confidence limits) and central estimates.

a. Has EPA identified and described the key sources of uncertainty in assessing cancer risks posed by tetrachloroethylene?
b. Is this analysis transparent and presented at a suitable level of detail for the IRIS assessment?
c. Does the assessment clearly and objectively present the choices made in developing reasonable upper-bound estimates of cancer risk posed by tetrachloroethylene?
d. The assessment includes tabular presentations of point-of-departure-based analyses that use different end points and approaches (see Tables 6-2, 6-3, 6-4, and 6-5). Is the information clearly presented and appropriately characterized?
e. In Section 6.2.2.2, the assessment presents exploratory calculations of potential probabilities of tumor response at low dose by using different functional forms. Is this analysis clearly presented and appropriately characterized?
f. Please discuss research subjects likely to characterize uncertainties better in future tetrachloroethylene cancer risk assessments.

Choice of Dose Metrics for Various Toxic Outcomes, PBPK Modeling, and Interspecies Scaling Approaches

Exposure to tetrachloroethylene results in the production of several metabolic products. The parent compound is used as the dose metric for neurotoxic effects, and the rate of formation of total metabolites in humans is used for cancer effects. Metabolite formation was modeled by using three PBPK models, which led to a range of cancer risk factors.

1. Please comment on the PBPK application for route-to-route extrapolation in developing an RfD and an oral slope factor from studies of inhalation exposure.

2. Please comment on the sufficiency of the available data to identify whether the parent compound or specific metabolites are responsible for the induction of cancer through tetrachloroethylene exposure.

3. Has EPA clearly and objectively presented
   a. Choice of dose metrics for different outcomes and their use in PBPK models?
   b. Strengths and weaknesses of different modeling approaches?
   c. The approach used in deriving the toxicologically equivalent human dose, including the application of an interspecies scaling factor ($BW^{3/4}$) to the fraction of the administered rodent dose that is metabolized?

4. Is EPA's conclusion that there is not a strong basis for preferring any one PBPK model for use in the risk assessment soundly and transparently characterized?

COMMITTEE'S APPROACH

The committee reviewed the material presented in EPA's draft IRIS assessment for scientific soundness, balance, and transparency. By the nature of the charge, the focus was on parts of the document that were critical for determining neurotoxicity and cancer end points. The review included evaluation of some of the primary literature cited by EPA, its approaches to evaluating and modeling data, and options for performing qualitative and quantitative assessment of uncertainties. Public comments submitted to EPA and to the committee on the draft assessment were considered. The committee also held public meetings at which it had the opportunity to ask questions of EPA staff, to obtain input from invited speakers who were doing research on tetrachloroethylene or related scientific issues, and to hear from other interested parties.

To identify new studies that should be considered in EPA's IRIS assessment, the committee performed a literature search for papers published from July 2004 (the official cutoff for EPA's comprehensive literature search) to March 2009. For the purposes of its review, the committee restricted its searches to MEDLINE and EMBASE. MEDLINE is produced by the U.S. National Library of Medicine and covers over 5,200 biomedical journals published in the United States and over 80 foreign countries. EMBASE is produced by Elsevier Science and indexes over 4,800 journals with a focus on the international literature. A simple search for "tetrachloroethylene," its synonyms, and its Chemical Abstracts Service registry number was performed. Literature retrieval was limited to studies pertinent to the evaluation of adverse health effects, such as toxicology studies (including studies on toxicokinetics and mode of action) and epidemiology studies.

Other sources of information that the committee considered included compilations of toxicology and human health information from national and international agencies and organizations, such as the Agency for Toxic Substances and Disease Registry, the International Agency for Research on Cancer, the California Environmental Protection Agency, and the European Union. Relevant publications from the National Research Council and the Institute of Medicine were also consulted. The committee and staff examined the reference lists included in EPA's draft assessment, major epidemiologic studies, review articles, and major compilations for relevant citations. Smaller targeted literature searches were performed to identify pertinent older literature and papers on specific topics and to gather general background information.

CONSIDERATION OF MODE OF ACTION

Much of the committee's task was focused on the mode of action or the toxic and carcinogenic effects of tetrachloroethylene. Because mode of action is considered throughout this report, a brief overview of what it means and of approaches to evaluating it is presented briefly here. The term *mode of action* is defined in the EPA cancer guidelines as a sequence of key events and processes, starting with interaction of an agent with a cell, proceeding through operational and anatomic changes, and resulting in cancer formation. A *key event* is an empirically observable precursor step that is itself a necessary element of the mode of action or is a biologically based marker of such an element. Mode of action is contrasted with *mechanism of action*, which implies a more detailed understanding and description of events, often at the molecular level, than is meant by *mode of action*.

The toxicokinetic processes that lead to formation of the active agent or its distribution to the target tissue, although considered in estimating dose, are not part of the mode of action as the term is used in the guidelines. Examples of possible modes of carcinogenic action are also presented in the guidelines, which state that they include mutagenicity, mitogenesis, inhibition of cell death, cytotoxicity with reparative cell proliferation, and immune suppression.

Understanding of mode of action is crucial for identifying susceptible life stages and determining appropriate approaches to extrapolation beyond the observable dose-response relationships. As a default, dose-response analysis for chemicals whose modes of action are expected to involve mutation involves linear extrapolation. Other modes of action may be modeled with either linear or nonlinear approaches after a rigorous analysis of available data under the guidance provided in the framework for mode-of-action analysis.

In the last decade, a continually evolving framework for considering weight of evidence for hypothesized modes of action and their human relevance has been developed and widely incorporated in guidance and risk assessments for individual chemicals by national and international agencies, including EPA. The framework is relevant to consideration of mechanistic data on both cancer and noncancer effects and sets the stage for informing dose-response relationships through consideration of hypothesized modes of action in the context of key events and their relevance to humans (for example, see Meek 2008). A framework requires delineation of a hypothesis with specified key events and then consideration of the weight of evidence of the hypothesized mode of action in animals in the context of such criteria as consistency, specificity, and biologic plausibility. Human relevance is then taken into account on the basis of consideration of the broader database and such matters as anatomy, physiologic variations, and human disease states.

Recent broad-based acceptance of mode of action and human relevance analyses is a function principally of their value in providing a structured approach to articulation of clear hypotheses, to description of the weight of evidence on which conclusions are based in the context of explicitly stated criteria, and to delineation of inherent uncertainties. The framework analyses ensure rigor in supporting and communicating the outcome of risk assessment and in facilitating the direction of resources to research to fill critical data gaps. The transparency promoted by framework analyses is expected to contribute to increased consistency in decision-making regarding modes of induction of cancer and later implications for dose-response analysis.

Mode-of-action analyses are based on the assumption that tumors in a single tissue are induced by a single mode of action, although in early stages several (seemingly competing) pathways may contribute. Mode of action is increasingly considered to incorporate toxicokinetics because often the critical first key event (which can be rate-limiting in the context of dose-response relationships) is activation to a toxic metabolite.

ORGANIZATION OF COMMITTEE'S REPORT

In the following chapters, the committee evaluates EPA's presentation and evaluation of the potential adverse health effects of exposure to tetrachloroethylene. Chapter 2 provides a brief overview of the toxicokinetics of tetrachloroethylene because understanding how the body handles tetrachloroethylene is critical for understanding its effects in the later chapters focused on specific organ systems. Chapter 3 presents an evaluation of the neurotoxic effects of tetrachloroethyelene; such effects were the basis of EPA's derivation of the RfC and RfD for tetrachloroetheylene, so the review focuses on evaluating the strengths and weaknesses of available studies and their utility in deriving reference values. Chapter 4 reviews EPA's presentation of the reproductive and developmental toxicity of tetrachloroethylene. That is followed by a chapter on the genotoxicity of tetrachloroethylene, which factors into the consideration of cancers of the liver (Chapter 6), kidney (Chapter 7), hematopoietic system (Chapter 8), and other organs (Chapter 9). Those toxicology reviews are followed by an assessment of EPA's derivation of the noncancer reference values (Chapter 10) and cancer-risk values (Chapter 11). Chapter 12 provides the committee's recommendations for future reassessments of tetrachloroethylene.

2

Overview of the Toxicokinetics of Tetrachloroethylene

It is important to be familiar with the toxicokinetics of tetrachloroethylene when evaluating the Environmental Protection Agency's draft Integrated Risk Information System (IRIS) assessment because many of the chemical's effects are thought to be associated with metabolites rather than with tetrachloroethylene itself. The draft IRIS assessment includes a thorough cataloging of the published literature on tetrachloroethylene metabolism, including consideration of the specific metabolite isoforms that may be involved and polymorphic variants. This chapter presents a brief overview of the absorption, distribution, metabolism, and excretion of tetrachloroethylene to provide context for discussions in this report. More specific toxicokinetic issues associated with specific outcomes and the committee's review of how they are handled in the draft IRIS assessment are discussed in later chapters.

Tetrachloroethylene is a volatile, lipophilic small molecule that is rapidly and extensively absorbed after inhalation and oral exposure. It can also be rapidly absorbed through the skin (Stewart and Dodd 1964), but dermal absorption appears to be a less important route of exposure. In humans, inhalation exposure to tetrachloroethylene typically results, within a few hours of exposure, in a pseudoequilibrium between inspired air and blood although there can be substantial interindividual differences in absorption behavior (Chiu et al. 2007). After oral dosing in animals, peak blood tetrachloroethylene concentrations are typically reached within 15-30 min, and systemic bioavailability is typically greater than 80% (Dallas et al. 1995); once absorbed, tetrachloroethylene is rapidly distributed throughout the body, and well-perfused tissues reach a pseudoequilibrium with blood within a few minutes. For example, after oral administration of a 10-mg/kg dose of tetrachloroethylene in rats, peak tissue concentrations occurred within 10-15 min in blood, brain, heart, lungs, kidneys, and liver (Dallas et al 1994). The elimination half-life of tetrachloroethylene was comparable among those tissues, between 6 and 7 hours (Dallas et al 1994). In poorly perfused tissues, such as fat and muscle, peak tetrachloroethylene concentrations are reached after a longer delay, which may be an hour or more than a day for adipose tissue. The elimination of tetrachloroethylene from fat is also much slower than that from other tissues and can take twice as long (Dallas et al. 1994). Because of its lipophilicity, the highest concentrations of tetrachloroethylene are found in adipose tissue (Savolainen et al. 1977; Dallas et al. 1994). In humans, the fat-to-blood concentration ratio has been estimated to be as high as 90:1 (Monster et al. 1979). Relatively high concentrations are also observed in the liver and brain (Savolainen et al. 1977). On the basis of animal studies and sparse human data, the brain concentration of tetrachloroethylene is 4-8 times the blood concentration (Dallas et al. 1994; Lukaszewski 1979).

The disposition of an absorbed dose of tetrachloroethylene occurs primarily through pulmonary excretion; metabolism is less important than for other chlorinated solvents, such as trichloroethylene. Mass-balance studies in rats with ^{14}C-labeled tetrachloroethylene indicated that 70% or more of an oral or inhaled dose can be recovered in expired air as the parent compound (Pegg et al. 1979; Frantz and Watanabe 1983). The next most important excreted fraction occurs in urine and feces, which may collectively account for up to 23% of an administered dose. A small portion of the dose (less than 3%) may be converted to CO_2 and exhaled. Most of the radioactivity recovered in urine can be attributed to formation of trichloroacetic acid, a nonvolatile metabolite of tetrachloroethylene that is excreted primarily in urine. That general pattern of disposition of tetrachloroethylene appears to be consistent after both oral and inhalation dosing (Pegg et al. 1979). However, it is important to note that the highest urinary and fecal elimination coincide with lower administered doses of tetrachloroethylene.

Despite the low overall metabolism of tetrachloroethylene compared with other chlorinated solvents, its metabolism has been studied extensively in both human volunteers and laboratory animals, using both in vivo and in vitro techniques. The studies showed that many metabolites are produced, including some known to be cytotoxic, mutagenic or both. Tetrachloroethylene metabolism can be viewed as having three pathways. The first is cytochrome P-450-mediated (CYP-mediated) oxidation. The second and third share a starting point: direct conjugation with glutathione to *S*-(1,2,2-trichlorovinyl)glutathione (TCVG) and then further processing to *S*-(1,2,2-trichlorovinyl)-L-cysteine (TCVC). For the second pathway, β-lyase catalyzes the formation of reactive products from TCVC. The third pathway is independent of β-lyase: TCVC is processed further by acetylation and sulfoxidation reactions. Genotoxic and cytotoxic metabolites are formed by each of these pathways. The predominant metabolic pathway is the CYP path, followed by the β-lyase pathway and then the β-lyase independent pathway. The TCVC derivatives are toxicologically important but quantitatively minor metabolites. A simplified scheme is shown in Figure 2-1.

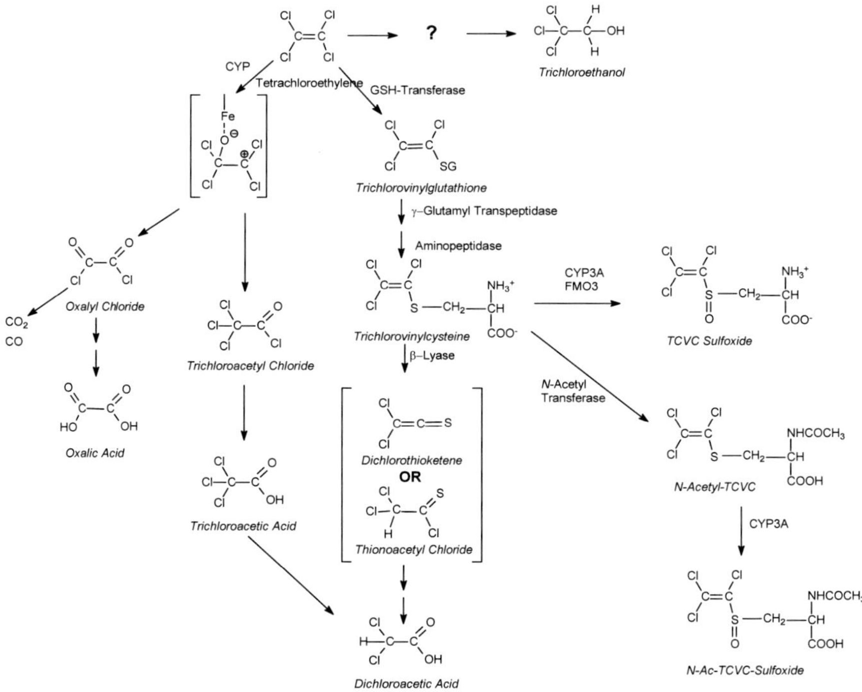

CYP Pathway β-Lyase Pathway β-Lyase-Independent Pathway

FIGURE 2-1 Simplified illustration of the metabolic pathways of tetrachloroethylene.

THE CYTOCHROME P-450 PATHWAY

The two major products of tetrachloroethylene metabolism by the CYP pathway are trichloroacetyl chloride and oxalyl chloride (Yoshioka et al. 2002). Trichloroacetyl chloride is mutagenic in the Ames test (DeMarini et al. 1994). Trichloroacetyl chloride reacts with lysine on protein to form stable trichloro adducts that can be detected with a specific antibody (Pahler et al. 1998). Trichloroacetyl chloride hydrolyzes to trichloroacetic acid (TCA), which produces liver cancer in mice (Nagano et al. 1998). Oxalyl chloride forms oxalic acid (possibly via oxalyl phosphate) or decomposes to CO_2 and CO. Oxalic acid has long been known to be nephrotoxic; calcium oxalate complexes result in tubular toxicity (Guo and McMartin 2005) and nephrolithiasis (Bushinsky et al. 2008).

Mechanistic studies on the products of CYP oxidation of tetrachloroethylene indicate that trichloroacetyl chloride is the predominant product of the CYP-tetrachloroethylene complex; formation of tetrachloroethylene epoxide is much less favored (Yoshioka et al. 2002). Formation of chloral by rearrangement of tetrachloroethylene epoxide has been postulated, as a pathway to trichloroethanol in analogy with trichloroethylene. Neither chloral nor chloral hydrate has been identified after tetrachloroethylene exposure. Chloral is a product of trichloroethylene oxidation by CYP although not through an epoxide intermediate (Miller and Guengerich 1982). Chlorine migration of the CYP-oxygenated trichloroethylene results in formation of chloral, whereas the product of tetrachloroethylene is trichloroacetyl chloride.

Rats and mice given tetrachloroethylene by gavage were reported to excrete trichloroethanol in urine (Dekant et al. 1986a). The formation of trichloroethanol from tetrachloroethylene has been reported after occupational exposure (Birner et al. 1996), but it was not confirmed in human volunteers exposed to tetrachloroethylene (Volkel et al. 1998; Chiu et al. 2007). Birner et al. (1996) noted that—on the basis of studies by Larson and Bull (1992)—TCA does not undergo reduction to trichloroethanol and could not explain trichloroethanol formation; a later publication from the same group concluded that trichloroethanol was an artifact of trichloroethylene exposure (Volkel et al. 1998).

Small amounts of dichloroacetic acid (DCA) may be produced by dechlorination of TCA (Larson and Bull 1992), but most DCA arises from the β-lyase pathway (Volkel et al. 1998; Dekant et al. 1988).

THE β-LYASE PATHWAY

Tetrachloroethylene is conjugated with glutathione to *S*-(1,2,2-trichlorovinyl) glutathione and is later processed by γ-glutamyl transpeptidase and aminopeptidase to TCVC (see Anders et al. 1988; Lash and Parker 2001). γ-Glutamyl transpeptidase is a brush-border enzyme that is found primarily in the renal proximal tubule and to a lesser extent in the bile canalicular membrane. B-Lyase forms 1-mercapto-1,2,2-trichloroethene, which can tautomerize to dichlorothionacetyl chloride or lose HCl to form dichlorothioketene. Dichloro-thionacetyl chloride and dichlorothioketene both yield dichloroacetic acid (Dekant et al. 1988). Dichlorothioketene reacts with lysine on protein to form stable dichloro adducts that can be detected with a specific antibody (Pahler et al. 1998).

Genotoxicity by the β-lyase pathway is supported by several studies. TCVG induces unscheduled DNA synthesis in mammalian kidney cells, and this response is blocked by inhibiting γ-glutamyltranspeptidase or β-lyase; such inhibition indicates that the genotoxic metabolite arises by the β-lyase pathway (Vamvakas et al. 1989a). The dichlorothioketene adenine and cytosine adducts, formed *in vitro* in organic solvents, do have stability under physiologic conditions and are potential mutagens (Muller et al. 1998a). The chlorofluoro analogue forms adducts with calf-thymus DNA and produces strand breaks. That analogue has chemical properties similar to those of dichlorothioketene; ^{19}Fl was substituted for a Cl to increase the sensitivity of detection (Muller et al. 1998b).

TCVC is cytotoxic to proximal tubule cells (Vamvakas et al. 1989b; McGoldrick et al. 2003). The toxicity is decreased by inhibition of β-lyase with aminooxyacetic acid. Elfarra and Krause (2007) reported potentiation of TCVC toxicity in rats by aminooxyacetic acid, which provides evidence for a β-lyase-independent mechanism in TCVC toxicity in rats in vivo.

Dichloroacetate is produced primarily through the β-lyase pathway and produces liver cancer in rats.

THE β-LYASE-INDEPENDENT PATHWAY

TCVC undergoes acetylation to its mercapturate *N*-acetyl-TCVC and then sulfoxidation to *N*-acetyl-*S*-(1,2,2-trichlorovinyl)-L-cysteine (*N*-Ac-TCVCS), which is mediated by CYP3A or flavin-containing monooxygenase (FMO). In addition, TCVC undergoes sulfoxidation to TCVC-sulfoxide (TCVCS); this is also mediated by CYP3A or FMO (Ripp et al. 1997).

TCVCS is a more potent nephrotoxicant than TCVC *in* vivo (Elfarra and Krause 2007). TCVC toxicity is increased by inhibition of β-lyase with aminooxyacetic acid (Elfarra and Krause 2007), underscoring the importance of the β-lyase-independent pathway for kidney toxicity. TCVCS mutagenicity appears to be untested. *N*-Acetyl-TCVC is not mutagenic in the Ames test but is more cytotoxic than *N*-acetyl-TCVC, which is mutagenic in the Ames test (Werner et al. 1996).

SPECIES DIFFERENCES

There are important differences between species in the metabolism and toxicity of tetrachloroethylene. Much work has focused on differences between humans and rats, particularly on differences that would influence the human risk of renal cancer that has been observed in rat bioassays. Comparison studies between rats and humans indicate that humans metabolize tetrachloroethylene less than rats; this is based on measurement of metabolites (Birner et al. 1996; Volkel et al. 1998) and on the formation of adducts that are detected by antibodies that are specific for either the CYP-derived trichloro adduct or the dichlorothioketene-derived dichloro adduct (Pahler et al. 1998).

The CYP Pathway

The CYP pathway is the predominant route of tetrachloroethylene metabolism in rats and humans. Plasma albumin adducted with the trichloro derivative, indicating metabolism by the CYP pathway, was found in rats and humans exposed to tetrachloroethylene at 40 ppm for 6 hours. Immunochemical staining was used; the staining of protein from rats was 15-20 times more intense than that of protein from humans (Pahler et al. 1999). Cumulative excretion of TCA in urine was measured in rats and humans after similar controlled exposure to tetrachloroethylene at occupationally relevant concentrations (Volkel et al. 1998). The committee used that data to calculate the ratio of urinary TCA excretion corrected for body mass in rats and humans. TCA excretion by rats was about 23 fold that of humans; or humans excreted about 4.4% of the amount excreted by rats.

The β-Lyase Pathway

Metabolism by the β-lyase pathway results in formation of dichloro protein adducts and DCA. Dichloro albumin adducts were detected in rat, but not human, blood samples after tetrachloroethylene exposure (Pahler et al. 1999). Even after immunoaffinity-column enrichment, the dichloro adduct was not detected in human samples. DCA is a stable product of the β-lyase pathway and is excreted in urine. Rats excreted DCA in urine at about one-tenth the amount of TCA, but DCA was not detected in urine collected from human volunteers after exposure to tetrachloroethylene (Volkel et al. 1998). That outcome is consistent with the lower activity of β-lyase in humans (McGoldrick et al. 2003).

The β-Lyase-Independent Pathway

Protein adducts resulting from the β-lyase-independent pathway have not been reported. *N*-Acetyl-TCVC, the mercapturate, is excreted in urine. Volkel et al. (1998) also measured urinary excretion of *N*-acetyl-TCVC after similar exposure to occupationally relevant concentrations of tetrachloroethylene. The Committee calculated the ratio of cumulative urinary excretion of *N*-acetyl-TCVC by rats to be about 5.5 fold that of humans; or humans excreted about 20% of the amount of *N*-acetyl-TCVC excreted by rats. Both rats and humans excrete much more TCA, the CYP-pathway product, than *N*-Ac-TCVC, but the ratio of *N*-acetyl-TCVC to TCA in humans is about 5 fold that of rats. That is, humans excrete relatively more tetrachloroethylene metabolites as *N*-Ac-TCVC than rats. That, too, is consistent with the lower activity of β-lyase in humans (McGoldrick et al. 2003); relatively more TCVC is metablized by the β-lyase-independent pathway in humans.

3

Neurotoxicity

This chapter reviews information presented in the Environmental Protection Agency (EPA) draft Integrated Risk Information System (IRIS) assessment of the effects of tetrachloroethylene on the nervous system. It considers first the human evidence, including an evaluation of EPA's selection of the most critical study on which to base its reference values, and then the evidence from experimental animal studies. The implications of the committee's evaluation on the derivation of EPA's reference values for tetrachloroethylene are discussed in Chapter 10.

HUMAN STUDIES

The epidemiologic studies available for evaluating the neurotoxic effects of tetrachloroethylene were generally cross-sectional. Only one study (Gobba et al. 1998) had outcome measures at two times. Although the cross-sectional study design is limited in establishing temporality in a causal association, the combination of the results of such studies with other information can help to establish an exposure-effect relationship.

In evaluating the human evidence, the committee applied several criteria for determining which studies were the most useful in establishing a reference concentration (RfC) for tetrachloroethylene. The criteria included three general characteristics: the validity of individual studies, the internal consistency of results (for example, Is there an association in the low-exposure group but not in the high-exposure group?), and the consistency of the findings with what is known from other sources (how the study fits into the overall picture of what is known). In selecting studies, the committee considered the target population, the study population, potential confounders, and possible selection or information biases. Statistical issues were also considered. Each study was looked at in the light of those factors, and studies were neither chosen nor rejected on the basis of their results. The selection criteria included consideration of the following factors and questions:

- **Populations:** Are the target and study populations well defined and described? Is the referent group representative of either the unexposed population (in a cross-sectional or cohort design) or of the source population (in a case-control design)? Studies with an inappropriate referent population were given less weight.
- **Selection of participants:** Are the methods for recruiting and enrolling study participants well described? Is there evidence of selection bias? If so, have the authors provided information on the magnitude of the bias? Whether an "effect" is observed in the exposed group is strongly influenced by the choice of the comparison or control group. Thus, the selection and composition of the comparison group is extremely important and in part determines the internal validity of the study. In some cases, there were clear selection biases (for example, selecting comparison groups for the exposed group that did not represent the counterfactual example). That introduces the possibility of selection biases that could easily create the appearance of differences, especially subtle ones, when differences do not exist.
- **Exposure assessment:** How well do the measurements used characterize tetrachloroethylene exposure? How are exposure groups defined? If individual exposure data were available, were they used, or was assignment to exposure groups based on ecologic criteria? In most cases, exposure was estimated at the time of a study. If it is assumed that exposure has only acute, reversible effects, cross-sectional studies are more appropriate. However, if occurrence of an effect when exposure concentrations are low requires long-term exposure, it is important to consider past exposure as well. Exposure assessment ranged from biologic measurements of tetrachloroethylene exposure to environmental exposure assessments. Studies that included measurements and analyses of exposure at the individual level were given greater weight.
- **Assessment of neurologic outcomes:** The end points that were measured in terms of relevance to the visual system and the degree to which the measures are influenced by cognitive function were considered. Studies that used less sensitive measures were given less weight, as were studies that used outcome measures that were more susceptible to observer bias or potential individual confounders (such as ability to follow instructions).
- **Confounding:** Observational studies are always subject to confounding when the exposed and referent groups are imbalanced with respect to factors that are not a result of the exposure but that are also related to the outcome. The committee considered the potential for differences in age, education, learning disabilities, and other variables to confound associations. If the potential for confounding was present and the effects of the confounding were not addressed by the study design or analytic methods, the results of the study were considered to be less credible.
- **Statistical analysis:** Statistical issues were considered, particularly whether the sample size was adequate and whether the approach to analysis was appropriate. Did the studies provide adequate information about the distribution

of exposure levels or results of outcome testing? Were the results influenced by only a few extreme values? If so, was that considered? If continuous data were available, were they used or collapsed as a binary variable, making dose-response analysis or assessment of thresholds impossible? Were tests for interaction of tetrachloroethylene exposure with other variables done? If so, were they properly interpreted?

Having applied those criteria, the committee disagreed with EPA's selection of the study by Altmann et al. (1995) as the critical study on the basis of which exposure limits should be estimated. EPA selected Altmann et al. (1995) because the data in it represent an environmental rather than an occupational exposure and because a standardized computer-assisted testing battery was used. Although those are reasonable considerations, they are not the most relevant for selecting a critical study. The committee concluded that the validity of the results of Altmann et al. (1995) was seriously compromised by the following methodologic deficiencies.

1. The reference group was inappropriate, because it did not represent the counterfactual example. The reference group included employees of the Public Health Office or the Medical Institution of Environmental Hygiene, none of whom resided at their place of employment and who may have lived outside the commercial city center. Personal characteristics as well as differences in exposures in the ambient environment may have confounded the analyses of exposure and neurobehavioral outcomes. Evidence of this selection bias is that although matched by age and sex, the referent group was clearly more educated than the exposed group. The distribution of the 14 exposed participants in the low, medium, and high education categories was four, eight, and two, respectively, and that of the 23 controls, one, 12, and 10. The effect of these differences on the study results could not be evaluated, however, because the numbers of years of education represented in the categories were not provided. Adjusting for education with broad categories rather than years of education is not adequate and can easily result in residual confounding by education. Evidence for residual confounding by education can be seen in the variability of results reported by Altmann et al. (1995) depending on the outcome measure. For example, no association between tetrachloroethylene and visual evoked potentials (VEPs) was found. That is important because changes in the visual system and abnormalities in VEPs have been associated with exposure to tetrachloroethylene and chemically related solvents (Bushnell and Crofton 1999; Gobba 2003; Bushnell et al. 2007; Benignus et al. 2009) and selected organic solvents (Benignus et al. 2009) and are unrelated to education. Measures of vigilance, attention, and visual memory are strongly associated with education and premorbid intelligence (Lezak et al. 2004). Those measures showed poorer performance in the exposed group, whereas measures of eye-hand coordination and finger tapping, which are weakly related to education and premorbid intelligence, were similar in the two groups.

2. The Neurobehavioral Evaluation System (NES) battery used to assess brain dysfunction related to exposure appropriately included four subtests that have been shown in other research to be associated with solvent exposure. However, the battery has no norms for this population, and some of the tests have not been well validated with regard to what they reveal about brain damage from any cause. The absence of norms makes it especially important to have standardized measures of intellectual function that can be used to characterize the native intellectual capacity of the two groups. Examples of such tests are the NES Vocabulary subtest, the Wide Range Achievement Test Reading subtest, and the Wechsler Adult Intelligence Scale Information subtest. Tests of native intellectual function like those are important to include in a battery used to assess neurocognitive outcomes because they are resistant to the effects of central nervous system insults from neurotoxic exposure. They can be used to control statistically for differences in premorbid function between exposed and control groups. Failure to use such measures can cause investigators to conclude that measured group differences in cognitive function are due to exposure when in reality they might exist without any exposure.

3. The authors indicated that there were 92 potentially eligible subjects, of whom 19 were selected as participants. It was unclear whether the 19 were selected because they were the only ones who had blood tetrachloroethylene over 2 μg/L, lived next to a dry-cleaning facility for at least 1 year, and had no occupational exposure to organic solvents. Even though a blood tetrachloroethylene concentration of over 2 μg/L was required for entry into the study, no concentrations were reported for five subjects (subjects 10-14) taken in their apartments (Figure 1A of Altmann et al. [1995]). Without those specifications, it is impossible to determine whether the sample was biased (that is, whether others were excluded for reasons other than study design).

4. Tetrachloroethylene was measured in air samples from homes for 7 days. Figure 1B of the paper purports to show indoor air concentrations for exposed participants and controls, but no concentrations are shown for the referent group. For subject 13 of the exposed group, there was no indoor air measurement, there was no tetrachloroethylene concentration in blood drawn in the apartment, and the blood concentration obtained at the time of testing was at the limit of detection (0.5 μg/L). Duration of residence of the 14 exposed ranged from 1 to 30 years; only mean duration was reported, not median. Given only a mean value, there is no way to know whether most of the exposed subjects had relatively short exposures and just a few had long exposures. The amount of time that residents spent in their apartments is unknown. Time out of the apartments before neurobehavioral testing was unknown but was believed to account for the lower blood tetrachloroethylene concentrations before testing. Two exposed subjects had blood tetrachloroethylene concentrations at the limit of detection when tested, whereas the blood concentrations of subject 4 were 30 μg/L in the apartment and 200 μg/L at the time of testing.

5. In the analyses, exposure is defined by group membership (yes or no) rather than by individual markers of exposure, so a dose-effect relationship could not be assessed. As stated above, group differences in neurobehavioral performance were more likely to be related to residual confounding by education or pre-exposure intellectual capacity than to exposure.

Another paper cited in the draft IRIS assessment that associated environmental tetrachloroethylene exposure with visual-contrast sensitivity (VCS) dysfunction reported on a pilot study by Schreiber et al. (2002). The study also suffered from important methodologic problems that limit its usefulness, including the criteria used to select the exposed group, selection of a noncomparable referent group, and errors in analysis and interpretation. It has been suggested that the significant results reported by Schreiber et al. were influenced largely by two exposed children who had diagnoses of developmental disorders (Storm and Mazor 2004). The total sample in the study was 17, of whom four were children; when the children were excluded from analyses, no significant associations were observed. Given the cross-sectional design of the Schreiber et al. study, it cannot be determined whether exposure preceded the developmental disorders. The small sample makes results highly sensitive to a few observations.

The published papers that the committee judged to be more appropriate to use as a point of departure for derivation of the RfC and reference dose (RfD) were Echeverria et al. (1995), Cavalleri et al. (1994) in combination with Gobba et al. (1998) and Altmann et al. (1990). The reasons for the selections are given below.

Echeverria et al. (1995) conducted a well-designed study of the relationship between acute and cumulative tetrachloroethylene exposure in dry-cleaning shops in Detroit, Michigan, and performance on a neuropsychologic battery. There was no "unexposed" group, but the referent group (lowest exposed; mean air tetrachloroethylene concentrations, not greater than 11.4 ppm) was in the same cohort of dry-cleaning shops as the "exposed" group (mean air tetrachloroethylene concentrations, not greater than 40.8 ppm). Using an internal referent group reduced the potential for the types of selection bias present in many other studies. In the analyses, several potential confounders were considered, including, age, education, verbal skill, alcohol consumption, and prior intoxicant exposure. The authors used a stepwise selection procedure for adjustment, but it is not clear which variables were ultimately used. After adjustment for the covariates, performance on tests for Wechsler Memory Scale Visual Reproduction, NES Pattern Memory, and NES Pattern Recognition was significantly poorer in workers who had a high index of lifetime tetrachloroethylene exposure than in workers who had a low index of lifetime tetrachloroethylene exposure (Table 3-1). Estimated lifetime tetrachloroethylene exposure was positively associated with self-reported "tension" (on the Profile of Mood States) and inversely associated with NES Pattern Recognition scores. Subanalysis demonstrated some similarity in the test results affected by tetrachloroethylene and alcohol consumption: Visual Reproduction, Pattern Memory, and Pattern Recognition. This similarity underscores the importance of adjusting for alcohol use in analyses of effects of tetrachloroethylene. The study is not without limitations in that recruitment was influenced by the lowering of the permissible exposure limit from 50 ppm to 25 ppm and by owners' emphasizing the cost of such a change for relatively little effect on health status; therefore, only 23 of a potentially eligible 125 shops participated, for a total of 65 exposed workers.

Cavalleri et al. (1994) examined color-vision loss in 35 dry-cleaning workers in 12 small dry-cleaning shops in Modena, Italy, and in controls who had no solvent exposure and were matched by age, sex, alcohol use, and cigarette-smoking. Inclusion criteria were "apparently healthy," average daily alcohol intake under 50 g/day, smoking fewer than 30 cigarettes/day, and corrected visual acuity of at least 6/10. Color vision was evaluated with the Lanthony 15 Hue desaturated panel, which was repeated 10 times. Few exposed or control workers were able to perform the test without error. Results wereexpressed as a color-confusion index (CCI) with errors in blue-yellow color vision. Tests were performed monocularly, and the mean CCI for both eyes was used in the analyses, although CCI may be affected in only one eye after tetrachloroethylene exposure. Air tetrachloroethylene concentrations obtained with personal passive sampling for 1 day produced a mean time-weighted average (TWA) for dry-cleaners of 7.27 ± 8.19 ppm (range, 0.38-31.19 ppm). The mean CCI for the dry-cleaners was significantly higher (1.192 ± 0.133) than that of controls (1.089 ± 0.117). The statistically significant relationship between TWA of tetrachloroethylene exposure and CCI depended on two extreme values. CCI was not related to duration of exposure or to an integrated index of exposure; only current exposure was known, and there were no data on tetrachloroethylene concentrations in previous years. The study established the protocol and baseline for the Gobba et al. (1998) study 2 years later, which was of greater interest to the committee.

TABLE 3-1 Estimated Mean[a] Neuropsychologic Test Results by Lifetime Exposure to Tetrachloroethylene in Study by Echeverria et al. (1995)

	Exposure Group		
Test	Low (N = 24)	Moderate (N = 18)	High (N = 23)
Visual reproduction	9.4 ± 1.21	8.9 ± 1.24	8.08 ± 1.24
Pattern memory	10.51 ± 0.82	10.36 ± 0.75	9.70 ± 0.72
Pattern recognition	14.39 ± 0.49	13.97 ± 0.49	13.83 ± 0.70
Tetrachloroethylene concentration at testing, ppm	< 0.6	4.3–12.1	11.4–41.8

[a]Means adjusted for covariates ± standard deviation.

Gobba et al. (1998) re-examined 33 of the workers from the Cavalleri et al. study for color-vision loss after an interval of 2 years. This study was unique in that it examined the same workers at two times. Overall, tetrachloroethylene concentrations remained unchanged for the whole group, but 19 workers (group A) had exposure to significantly increased tetrachloroethylene concentrations at the time of the second assessment, and the remainder (group B) had exposure to significantly lower concentrations because of changes in the processes used in their dry-cleaning shops. Demographic information was provided on the group as a whole but not the two subgroups. The mean CCI increased significantly over the 2 years in group A (from 1.16 ± 0.15 to 1.26 ± 0.18) but remained unchanged in group B (1.15 ± 0.14 and 1.15 ± 0.13). In comparison, the control group from the Cavalleri et al. study, which was not re-examined in the Gobba et al. study, had a mean CCI of 1.08 ± 0.10. The clinical significance of these CCI changes is uncertain. The participants in the Gobba et al. study had exposure concentrations closer to those reported in environmental studies. That the CCI did not improve in the group with lower tetrachloroethylene exposure might be because improvement in workplace conditions had been in place for only a short time or because the visual changes are not reversible.

Altmann et al. (1990) randomly allocated 22 healthy young male subjects to exposure to tetrachloroethylene at 10 ppm or 50 ppm in a chamber for 4 hours on 4 consecutive days, and blood samples were taken for tetrachloroethylene testing and visual and neurophysiologic tests were performed. All subjects had normal visual acuity and no previous solvent exposure. Increased latency in VEPs was observed in subjects exposed to tetrachloroethylene at 50 ppm, and decreased latency at 10 ppm; the greatest effect was observed on the last day of exposure. VEPs with the smallest visual angle and on the last day of exposure provided the greatest intergroup differences. VCS tests on five subjects (two at 50 ppm and three at 10 ppm) showed improvement at the low and intermediate spatial frequencies in the 10-ppm group but loss in the 50-ppm group. Brainstem auditory evoked potentials were not associated with tetrachloroethylene exposure. The lowest observed-adverse-effect level (LOAEL) appeared to be 10 ppm for VEP outcomes.

A second paper (Altmann et al. 1992) published on the above study summarized data on neurobehavioral outcomes but is not recommended for use in determining reference values. Performance during 4 days of exposure was compared with performance obtained on day 1 in the chamber, when there was no exposure. The NES subtests measuring mood and "cognitive function" showed no decrement in performance with days of exposure, but the continuous performance test, tracking task (hand-eye coordination subtest), and simple reaction time task showed improvement over time that was more pronounced in the 10-ppm control group than in the 50- ppm exposure group. However, the measure of premorbid function used in the study (a vocabulary test) was not included as a control measure in the data analyses; it might have affected the outcomes on all NES subtests, especially those of learning and memory. Some NES subtests were given only twice and some at every session; it is not clear which were given when, but it might have influenced which test outcomes had significant results because of differences in practice effects.

ANIMAL STUDIES

This section describes controlled-exposure studies of experimental animals. As noted in the draft IRIS assessment, most animal studies have involved inhalation exposures to tetrachloroethylene at concentrations of about 30 ppm to over 1,000 ppm or administration by noninhalation routes of tetrachloroethylene at 100-to 4,000 mg/kg. Because of the relevance of the exposure regimen, the inhalation studies are emphasized here. However, it should be noted that studies like that of Warren et al. (1996) and Moser et al. (1995) deliver a known amount of tetracholorethylene by other routes (for example, by gavage) and also support tetrachloroethylene's neurotoxicity. Warren et al. reported effects on a refined end point, schedule-controlled behavior, and linked behavioral deficits to blood and brain concentrations. Moser et al. (1995) used a broad range of doses administered acutely or "sub-acutely" (14 days) and reported LOAELs and no-observed-adverse-effect levels (NOAELs) on a well-characterized Functional Observational Battery.

Incorporating the animal literature into an assessment of tetrachloroethylene's neurotoxicity has several advantages. The animal literature can demonstrate the plausibility of claims that neurotoxicity occurs, identify the role of dose and duration of exposure in neurotoxicity, discover neurotoxic effects for further study in humans, confirm with controlled exposures that neurotoxicity occurs in a specific domain, link effects to tissue concentrations, and determine mechanisms of action and similarities and differences between other compounds in the same class. The animal studies entail known histories and living conditions and controlled exposure conditions, usually over a range of doses or concentrations; this allows assessment of dose-effect relationships under conditions that are less influenced by the covariates and biases that hamper the interpretation of human exposures.

The literature describing controlled acute and subchronic inhalation exposures of laboratory animals is summarized in the EPA document. The end points affected include neurotransmitter or neurochemical concentrations (Honma et al. 1980; Nelson et al. 1979; Briving et al. 1986; Karlsson et al. 1987), long-chain fatty acid concentrations (Kyrklund et al. 1984, 1987), RNA expression (Savolainen et al. 1977), DNA expression and brain weight (Rosengren et al. 1986; Wang et al. 1993), electrophysiologic measures and evoked potentials (Mattsson et al. 1998), and locomotor activity (Savolainen et al. 1977; Kjellstrand et al. 1985; Szakmary et al. 1997), all of which indicate tetrachloroethylene's neurotoxcity. Some studies published after the draft IRIS assessment was written have applied physiologically based pharmacokinetic (PBPK) modeling to characterize not only the dose to which an animal is exposed but the concentration at the target tissue for neurotoxicity, the brain (e.g,, Boyes et al. 2009).

The incorporation of PBPK modeling will facilitate generalization among species and among routes of exposure. The process can contribute to the identification of mechanisms and modes of action and can enhance understanding of the comparative toxicity of different solvents.

The animal studies have limitations. Most notably, as in the studies of controlled human exposure, they use concentrations that are much higher and durations that are much shorter than those experienced environmentally or occupationally. Incorporating their results into a risk assessment must entail the application of uncertainty factors to identify hazard at environmentally, or even occupationally, relevant concentrations. In addition, the dependent measures in most studies differed from those identified in the human literature as particularly sensitive to tetrachloroethylene exposure. In contrast, recently published papers, such as those by Oshiro et al. (2008) and Boyes et al. (2009), use end points that are directly relevant to humans.

The draft IRIS assessment reviews two papers by Kjellstrand et al. (1984, 1985 [see Table 4-6, page 4-409 of EPA 2008]) for neurotoxicity. However, the 1984 study is not appropriate for assessing neurotoxicity; its strengths are that it involved doses that ranged from 9 to 3,600 ppm and durations that ranged from 1 to 120 days and continuous exposure or exposure for a different number of hours per day, but no central nervous system end points were examined. EPA reports that brain butyrylcholinesterase activity was affected, but plasma was analyzed, so the relevance to neurotoxicity is unclear. Some mice were examined for locomotor activity, but exposure and effects are poorly described and unusable. Although the exposure was acute, the relationship between locomotor activity and exposure is described better in the 1985 paper.

Overall, the animal studies support the conclusion that tetrachloroethylene is neurotoxic, but, except for the study by Mattsson et al. (1998), the end points used in the animal studies that were reviewed by EPA were nonspecific and not directly related to the visual or cognitive effects reported in the human literature. The studies therefore provide only indirect support for EPA's conclusions. The studies by Mattsson et al. entailed exposure 6 hours/day 5 days/week for 13 weeks and examined VEP and other functional effects, so their results are directly pertinent to human exposures. A NOAEL and a LOAEL were identified. Several related reports have been published since the draft IRIS assessment was written (for example, Boyes et al. 2009; Oshiro et al. 2008); they describe dose-effect relationships, spanning a broad range of doses, between acute exposure and visual and signal-detection end points.

In the Boyes et al. (2009) study, rats were exposed head-only to tetrachloroethylene while VEPs were recorded. Exposures were to concentrations of tetrachloroethylene ranging from 1,000-4,000 ppm for 1-2 hours, using concentration and time combinations derived from kinetic analyses. The most sensitive end point was the F2 (frequency-doubling) component of the evoked potential spectrum, a measure thought to reflect the activity of cortical neurons that respond to both stimulus offset and onset. Boyes et al. also conducted a toxicokinetic analysis relating exposure concentration (250-4,000 ppm) and duration (1 hour followed by a 6-hour washout period) to brain concentration. From this analysis, the investigators were able to link brain concentrations of tetrachloroethylene to visual function and to estimate an ED_{10} of 0.68 mg/L and ED_{50} of 47 mg/L.

In the study by Oshiro et al. (2008), rats were exposed by inhalation to tetrachloroethylene at 500, 1,000, and 1,500 ppm for 1 hour, during which a visual signal detection task was performed. Rats were trained to indicate the occurrence or nonoccurrence of a light flash during a trial period that lasted from 0.3 to 24.39 seconds, and individual trial durations were random. Exposure to tetrachloroethylene did not change the number of "correct" detections, but significantly increased the number of times that the rats incorrectly indicated a signal (false alarm), increased response time, and decreased the number of trials completed. The false-alarm rate was affected at the lowest concentration (500 ppm) and a NOAEL was not identified. The authors concluded that the results suggest attention deficits.

EPA also reviewed animal studies conducted with intraperitoneal or oral exposure. The studies of exposure of adults included functional observational batteries (Moser et al. 1995), locomotor activity (Fredriksson et al. 1993; Motohashi et al. 1993), and schedule-controlled operant behavior (Warren et al. 1996). EPA did not use the studies in establishing an oral RfD for chronic adult exposures, because effects occurred at high doses (150 mg/kg per day or higher) in the well-controlled studies.

The mode of action for tetrachloroethylene's neurotoxicity is discussed in a separate section of the draft IRIS assessment (Section 4.6.4). The assessment notes that while the mechanism by which tetrachlorethylene acts is unknown, the evidence is good that it acts on ligand-gated ion channels like other organic solvents. EPA correctly notes that solvents act similarly to ethanol on $GABA_A$ receptors and that there are orderly structure-activity relationships, but the citation in support of this observation (Mihic 1999) reviews ethanol and not other solvents. As implied in the IRIS assessment, tetrachloroethylene's effects on brain fatty acids are interesting but its functional significance is not clear. A weakness of the IRIS assessment's treatment of the evidence on tetrachloroethylene's mechanism of neuorotoxic action is that it is entirely descriptive and isolated from the rest of the document. Specifically, the implication that it resembles other volatile organic solvents is not used elsewhere in the document in support of tetrachloroethylene's toxicity to the adult or the developing nervous system. In light of the importance of neurotoxicity to the development of the RfC, this is surprising.

DEVELOPMENTAL NEUROTOXICITY

The literature on developmental neurotoxicity is limited. EPA's discussion of this important issue is distributed between the sections on neurotoxicity and reproductive toxicity. In light of the sensitivity of the developing nervous system to neurotoxicants, including solvents (Costa et al. 2004; Grandjean and Landrigan 2006; Slikker 1994), the topic should have been given separate treatment. The EPA document appropriately raises concerns that the studies of tetrachloroethylene-exposed children are small or sufficiently problematic that firm conclusions cannot be drawn from them. Several effects have been reported, including alterations in sensorimotor function (Nelson et al. 1979; Umezu et al. 1997), brain neurochemistry (Nelson et al. 1979), and locomotor activity (Fredriksson et al. 1993; Motohashi et al. 1993; Nelson et al. 1979; Szakmary et al. 1997). Some of these studies used very high concentrations, but others involved concentrations relevant to potential human exposures.

Nelson et al. (1979) exposed pregnant rats to tetrachloroethylene at 900 ppm on gestational days 7-13 or 14-20 or at 100 ppm on days 14-20. No significant tetrachloroethylene-related effects were reported in the animals exposed at 100 ppm, but effects were noted in those exposed at 900 ppm. The tetrachloroethylene-exposed dams consumed less feed and gained less weight than air-exposed controls. No significant differences in growth were noted in offspring, but the draft IRIS assessment incorrectly states that diminished weight gain in offspring was reported. Offspring showed deficits in neuromuscular and sensorimotor functions and increases in locomotor activity.

Fredriksson et al. (1993) also reported changes in locomotor activity in 60-day-old rats after oral exposure to tetrachloroethylene administered (at 5 and 320 mg/kg) on postnatal days 16-20; the effects were not dose-related. The draft IRIS assessment appropriately raised a concern about adequate control for litter effects in the study. It is widely accepted that litter effects must be controlled for in analyses of developmental exposure. Usually litter effects are handled by including only one pup, or one pup per sex, from each litter in studies of prenatal or perinatal exposures. That is, to avoid "litter effects," the litter should be the statistical unit. A failure to follow that convention inflates the type I error rate. Fredriksson et al. (1993) did not follow it but instead assigned pups to treatment groups randomly, so some treatment groups contained siblings. Some of the authors of the paper have argued that their approach is appropriate and does not inflate the type I error rate (Ericksson et al. 2005); their discussion is also cited in the draft IRIS assessment. Because exposures took place on postnatal days 16-20, the extent to which litter effects confounded the results in the 1993 Fredriksson et al. study is unclear. Nonetheless, the absence of a dose-effect relationship is of concern.

In a short communication, Kyrklund and Hagid (1991) described changes in brain fatty acids of neonatal guinea pigs exposed to tetrachloroethylene at 160 ppm during gestation, but the samples were very small, and many important details were lacking. As noted in the draft IRIS assessment, there was evidence of litter effects in this study, and EPA correctly notes that there are concerns about the absence of a dose-effect relationship and of important methodologic considerations, such as use of non-blinded observers on end points that involved subjective observations and difficulty in relating intraperitoneal routes of administration to oral or inhalation routes.

As noted in the draft IRIS assessment (section on "Mode of Action for Neurotoxic Effects" [4.6.4]), tetrachloroethylene has much in common with other volatile organic solvents, anesthetics, and alcohols. These shared mechanisms, coupled with similarities in the kinetics of these compounds and the high vulnerability of the developing brain to organic solvents and alcohols, raise concerns about the vulnerability of the developing organisms to tetrachloroethylene. The material on developmental neurotoxicity, while identifying the studies directly pertinent to tetrachloroethylene, omits mention of evidence that might be derived from similarly acting compounds. A separate section might have addressed these issues more thoroughly.

FINDINGS AND RECOMMENDATIONS

EPA's selection of neurotoxicity with emphasis on the outcomes of cognitive and visual dysfunction in adults is appropriate as an end point for deriving a point of departure for development of its reference values. However, the committee disagrees with EPA that the study by Altmann et al. (1995) should be the basis for the noncancer risk values. The committee recommends the use of studies by Altmann et al. (1990), Cavalleri et al. (1994) as a baseline for Gobba et al. (1998), and Echeverria et al. (1995). A new animal study by Boyes et al. (2009) also provides a strong basis for a point of departure. Those five studies provide a stronger scientific basis for deriving the RfC and RfD. Despite the importance of the developing nervous system, the literature on potential neurodevelopmental effects is not sufficient to support the derivation of an RfC. This does not mean that developmental neurotoxicity is unlikely. The broader solvent literature raises significant concern about potential developmental neurotoxicity. While the draft IRIS assessment notes that tetrachloroethylene enters the developing brain, it appears to dismiss the potential for developmental neurotoxicity independent of reproductive or maternal toxicity.

Additional research may help to fill gaps in the evidence. For example, studies of developmental neurotoxicity are needed to fill an important gap in the database on tetrachloroethylene. Well-designed epidemiology studies of tetrachloroethylene and neurological end points that characterize both past and current exposure would be helpful. These studies should be done in populations with a range of exposures (such as occupational studies with a wide distribution of exposure and environmental exposures via both air and water).

4

Reproductive and Developmental Effects

The Environmental Protection Agency (EPA) draft Integrated Risk Information System (IRIS) assessment describes the key animal developmental-toxicity and reproductive-toxicity studies of tetrachloroethylene in Section 4.7.2 and provides useful summaries of the study results in its Tables 4-8 and 4-10. In evaluating the studies described by EPA, the committee applied several criteria to determine whether there is sufficient evidence to identify tetrachloroethylene as a reproductive or developmental toxicant in animals and to identify a reference concentration based on reproductive or developmental end points. The criteria included consideration of identification of adverse effects that were not confounded by excessive maternal toxicity, use of multiple experimental exposures, identification of a no-observed-adverse-effect level (NOAEL), and conformity with current regulatory testing guidelines.

The committee agrees with the NOAEL of 100 ppm based on the study by Tinston (1994). EPA'a derivation of a comparative reference value (RfV) based on reproductive or developmental toxicity is an important addition to the toxicologic information on tetrachloroethylene and will be helpful in assessing potential health risks related to these end points. However, EPA's rationale for selecting the Tinston (1994) study instead of the Carney et al. (2006) study for the benchmark dose analysis and derivation of the RfV is not presented in the document and therefore is unclear. A major criticism of Section 4.7.2 has to do with the general lack of transparency regarding the critical analysis that EPA conducted of the studies described. The strengths and limitations of individual studies are not adequately discussed, and evaluations of reported maternal toxicity and comparisons of studies that yielded supporting or conflicting evidence of developmental or reproductive toxicity are not adequate. As a result, the reader cannot readily conclude that EPA had sufficient data for a risk assessment. Furthermore, the scientific basis for considering some studies and not others for derivation of a comparative RfV based on reproductive or developmental toxicity is not apparent. EPA does not state whether the experimental animal evidence of tetrachloroethylene-induced developmental toxicity and reproductive toxicity is sufficient or insufficient on the basis of criteria in its risk-assessment guidelines. Some of the specific deficiencies in Section 4.7.2 are described below.

LIMITATIONS OF THE DATABASE

Information analogous to that on page 4-124 of the draft IRIS assessment, which discusses general limitations of the human reproductive-toxicity and developmental-toxicity studies, would be useful. It would provide a context for the descriptions of individual studies and would be helpful in characterizing the animal developmental-toxicity and reproductive-toxicity data available for hazard identification and dose-response evaluation. For example, only two studies of the reproductive toxicity of tetrachloroethylene are described, and many of the developmental-toxicity studies described have limitations. The limitations include use of a single exposure level, insufficient study details, excessive maternal toxicity, and lack of conformity with current EPA and Organisation for Economic Co-operation and Development (OECD) regulatory testing guidelines because of when the studies were conducted.

COMBINED DISCUSSION OF REPRODUCTIVE AND DEVELOPMENTAL TOXICITY

EPA discusses the evidence on reproductive toxicity and developmental toxicity together. Without a separate discussion of each, it is difficult to identify conflicting data and data gaps and to assess whether there is sufficient evidence of toxicity for each end point according to the criteria in the EPA (1991, 1996) guidelines. The sequence or order in which the studies are described in Section 4.7.2 complicates the issue. The two studies that provide specific information on the reproductive toxicity of tetrachloroethylene, Tinston (1994) and Beliles et al. (1980), are not discussed sequentially. The end-point-specific evidence from the well-conducted Tinston (1994) reproduction study and the Carney et al. (2006) developmental-toxicity study is either not stated or not emphasized by EPA. For example, EPA does not conclude from the Tinston (1994) two-generation reproduction study that tetrachloroethylene had no significant effect on reproductive performance or fertility in rats at up to 1,000 ppm. The results of the Beliles et al. (1980) study, which showed that tetrachloroethylene at 500 ppm had no significant effect on the sperm of rats, are consistent with the adverse effect on fertility in the Tinston study, but the relationship of this finding to the Tinston (1994) study is not discussed. The Summary on page 4-134 does not mention the results of the Carney et al. (2006) developmental study, which showed that tetrachloroethylene at 249 ppm, in the absence of maternal toxicity, can produce developmental toxicity in rats (reduced fetal and placental weights and incomplete ossification of thoracic vertebral centra).

EVALUATION OF THE RELATIONSHIP OF MATERNAL AND DEVELOPMENTAL TOXICITY

The EPA risk-assessment guidelines (EPA 1991, p. 18) state: "Since the final risk assessment not only takes into account the potential hazard of an agent, but also the nature of the dose-response relationship, it is important that the relationship of maternal and developmental toxicity be evaluated and described." It is not clear whether EPA evaluated the range of maternal-toxicity data (mild to severe effects) that are reported in the studies described, inasmuch as interpretation of the data with regard to the developmental toxicity of tetrachloroethylene is not presented. For example, in the Schwetz et al. (1975) study, tetrachloroethylene produced a statistically significant increase in resorptions and mild, statistically significant maternal toxicity (4-5% reductions in mean maternal body weight compared with controls) in rats. Food consumption and liver weights were not affected by tetrachloroethylene exposure. Maternal toxicity is listed in the EPA draft's Table 4-8 as an "Effect," but there is no discussion of its relationship to the increased resorptions. According to the EPA risk-assessment guidelines, the increased resorptions in the Schwetz et al. (1975) study represent tetrachloroethylene-induced developmental toxicity in that they were produced at doses that caused minimal maternal toxicity. Maternal toxicity (decreased body weight gain and increased liver weight and serum enzyme activities) at tetrachloroethylene concentrations of 221, 664, and 1,254 ppm is also listed as an "Effect" in Table 4-8 for the Szakmary et al. (1997) study. EPA does not point out that the excessive maternal toxicity at 664 and 1,254 ppm (decreases of 37% and 40% in maternal body-weight gain, respectively, compared with 13% at 221 ppm) makes the developmental effects (such as skeletal retardation and decreased fetal weight) difficult to interpret and of limited value on the basis of its risk-assessment guidelines.

STUDY STRENGTHS AND LIMITATIONS AND CONSISTENCY OF RESULTS

Section 4.7.2 of the EPA draft does not identify the studies that are scientifically strong and the studies that are weak. Supportive and conflicting studies in the database also are not adequately identified. For example, EPA does not explain why confidence in the Tinston (1994) and Carney et al. (2006) studies should be higher than in the other studies described. In addition to being well conducted, both Tinston and Carney et al. have multiple experimental exposures, report effects associated with lower exposures that are not confounded by excessive maternal toxicity, and identify NOAELs. As indicated on page 5-4 of the draft, EPA considered those studies supportive of a point of departure to derive an RfV based on some of these strengths. EPA (2008, p. 4-137) indicates that reduced birth weight was found in five studies but does not discuss the consistent finding of tetrachloroethylene developmental toxicity at similar concentrations in Tinston (300 ppm), Carney et al. at (249 ppm), Schwetz et al. (300 ppm), and Szakmary et al. (221 ppm) or the conflicting finding of no developmental toxicity at 500 ppm in the Hardin et al. (1981) study. The limitations of Hardin et al. (single exposure level and lack of minimal maternal toxicity), Schwetz et al. (single exposure level), Nelson et al. (1979) (insufficient study details), and Szakmary et al. (lack of dose-response relationship because of excessive maternal toxicity at higher exposure levels) also are not discussed. In addition, the studies that do not conform to EPA and OECD regulatory testing guidelines are not identified.

STRENGTH OF EVIDENCE

The summary of the data on the developmental toxicity of tetrachloroethylene from selected studies is not particularly helpful, because EPA did not present its evaluation of the information and the basis for citing particular studies and study results is unclear. For example, EPA cites limited developmental-toxicity studies, such as Szakmary et al. (1997) and Schwetz et al. (1975), but does not cite Carney et al. (2006), the strongest one. EPA's reason for citing tetrachloroethylene-induced behavioral changes as evidence of developmental toxicity in the summary also is not clear, and the citation does not seem to be supported by the data. Tetrachloroethylene's effects at 1,000 ppm in the Tinston (1994) study are described on page 4-131 as central nervous system (CNS) depression and in Table 4-9 as behavioral effects. CNS depression appears to be more accurate on the basis of the symptoms described. The behavioral effects reported by Szakmary et al. (1997) are confounded by excessive maternal toxicity, and tetrachloroethylene had minimal effects on the behavior of rats in the study by Nelson et al. (1979). EPA provides no summary information on the reproductive toxicity of tetrachloroethylene even though data are available from a well-conducted two-generation reproduction study (Tinston 1994). Stating whether tetrachloroethylene can be identified as a developmental toxicant or a reproductive toxicant according to the criteria in the EPA developmental-toxicity risk-assessment guidelines (EPA 1991) and reproductive-toxicity risk-assessment guidelines (EPA 1996) would be helpful to risk managers and others and would help to identify data gaps.

For example, *there is sufficient evidence to identify tetrachloroethylene as a developmental toxicant in experimental animals* on the basis of the results of Carney et al. (2006) and Tinston (1994). That conclusion is consistent with the developmental-toxicity risk-assessment guidelines (EPA 1996, p. 40), which state: "The minimum evidence necessary to judge that a potential hazard exists generally would be data demonstrating an adverse developmental effect in a single, appropriate, well-conducted study in a single experimental animal species." *There is insufficient evidence to indicate that tetrachloroethylene does not cause reproductive toxicity in experimental animals* on the basis of the negative findings on reproductive performance and fertility in Tinston. That conclusion is consistent with the reproductive-toxicity risk-assessment guidelines (EPA 1991, p. 72), which state: "The minimum evidence needed to determine that a potential hazard does not exist would include data on an adequate array of endpoints from more than one study with two species that showed no adverse reproductive effects at doses that were minimally toxic in terms of inducing an adverse effect. Information on pharmacokinetics, mechanisms, or known properties of the chemical class may also strengthen the evidence."

ATTRIBUTING DEVELOPMENTAL TOXICITY TO TRICHLOROACETIC ACID

EPA's speculation in Section 4.7.4 of the draft that trichloroacetic acid (TCA) is the causative agent in the developmental toxicity of tetrachloroethylene does not seem scientifically sound, and the discussion is not balanced. The available scientific data appear to contradict EPA's speculation. In the studies by Schwetz et al. (1975) and Carney et al. (2006), trichloroethylene (in contrast with tetrachloroethylene) did not cause developmental toxicity even though higher concentrations of TCA should have been produced from trichloroethylene than from tetrachloroethylene. In addition, tetrachloroethylene and TCA produce different types of developmental toxicity. Oral administration of TCA has consistently produced cardiac malformations in rats (Smith et al. 1989; Johnson et al. 1998). Dichloroacetic acid (DCA) also produces cardiac malformations when administered orally to rats (Smith et al. 1992; Epstein et al. 1992). The malformations produced by TCA and DCA are consistent with the teratogenic potential of other weak acids, such as valproic acid and ethylhexanoic acid (Scott et al. 1994), but are not consistent with tetrachloroethylene-induced developmental toxicity. The developmental toxicity produced by tetrachloroethylene did not include cardiac malformations in any of the studies described by EPA in Section 4.7.2. EPA's discussion of the evidence supporting TCA as the causative agent in tetrachloroethylene developmental toxicity is not balanced. EPA did not comment on the relatively high concentrations of TCA required to cause developmental toxicity compared with the concentration expected to result from metabolism of tetrachloroethylene in vivo or on whether this could account for the difference in the type of developmental effects that result from tetrachloroethylene exposure. The lack of information on the availability of metabolized TCA to the developing fetus and the potential differences related to oral vs inhalation exposure in the TCA and tetrachloroethylene studies, respectively, also were not addressed.

EPIDEMIOLOGIC STUDIES

Few epidemiologic studies bear on possible associations between exposure to tetrachloroethylene and the specific adverse reproductive outcomes considered. Most of the available studies have serious methodologic limitations and so are not particularly informative as to the potential adverse reproductive effects of tetrachloroethylene exposure. Challenges that commonly confront investigators conducting epidemiologic studies of environmental determinants of reproductive health were evident in the available literature, specifically, standard case definitions, systematic ascertainment of end points, correct classification of exposure with respect to timing of pregnancy, and specificity of exposure to tetrachloroethylene.

The draft IRIS assessment considered the evidence on reproductive effects of tetrachloroethylene to be limited but cited spontaneous abortion as the outcome for which the evidence of an association with tetrachloroethylene was strongest on the basis of results in three papers (Kyyronen et al. 1989; Olsen et al. 1990; Doyle et al. 1997). In general, the committee agrees with EPA's assessement but takes a cautious view of inferences about the reproductive effects of tetrachloroethylene. The committee considered the work by Doyle et al. (1997) and Kyyronen et al. (1989) to be the most methodologically sound because they were based on cohorts of employed women about whom there was some information on tetrachloroethylen exposure and there was adequate evidence that the spontaneous abortions were validly reported. The studies examined spontaneous abortion in recognized pregnancies in cohorts of dry-cleaning and laundry workers; both reported an increased risk of spontaneous abortion in women who worked in dry-cleaning while pregnant. Nevertheless, both studies were limited by potential selection bias and small sample sizes and did not adequately address early fetal loss. They provide limited but supportive evidence of an association between tetrachloroethylene exposure and spontaneous abortion. The other study that EPA found compelling was that by Olsen et al. (1990); this study, although methodologically sound, was limited by the small number of events in the exposed groups.

There was also limited evidence of effects of tetrachloroethylene exposure on the developing fetus in a well-designed study from Camp Lejeune, North Carolina (Sonnenfeld et al. 2001). An increase in small-for-gestational-age cases was observed in children born to older women and women who had a history of fetal loss, but little effect was observed in other segments of the population. That discrepancy was difficult to resolve and may be spurious. (After publication of this study, it was discovered that some members of the control population were misclassified and were actually exposed, so the analyses in the paper are no longer valid.) EPA is inconsistent in characterizing the strongest evidence of reproductive toxicity. In "Characterization of Hazard and Dose Response" (EPA 2008; Section 6.1.3, page 6-5, lines 5-6), EPA cites "some evidence for growth retardation in infants born to mothers residing in housing with drinking water contaminated with tetrachloroethylene" as the main evidence of a reproductive outcome of concern. That conflicts with the conclusions in Chapter 4, where EPA indicates that the strongest evidence is on spontaneous abortion on the basis of the occupational studies.

EPA also considered potential male-mediated effects of tetrachloroethylene (Eskenazi et al. 1991a,b). Semen-analysis measures in dry-cleaning and laundry workers were compared. The reported differences were subtle and did not always favor the exposed or unexposed. The second study examined total fertility in the wives of dry-cleaners by using standardized fertility ratios; this study was uninformative in that it was too small to evaluate fertility patterns.

In general, the committee did not consider the draft section on adverse reproductive and developmental outcomes to be balanced in the presentation or critique of studies. The committee's general impression was that the section focused primarily on studies that reported results that confirmed a positive association and that the effect of methodologic limitations of the studies on the validity of results was not fully appreciated. For example, in discussing possible reasons for failure to find associations between tetrachloroethylene exposure and adverse outcomes (page 4-121, line 33, through page 4-122, line 7), the draft did not consider the possibility that there is no association. In another case, the draft assessment refers to a "strong but imprecise association between IUGR [intrauterine growth restriction] and exposure to tetrachloroethylene (OR =12.5, 95% CI not given" (page 4-122, lines 8-12), but this result is based on *a single exposed case*. EPA's description suggests an impressive finding. A more appropriate discussion would have stated there were too few exposed cases to calculate a measure of association reliably and would not have cited the odds ratio.

In addition, the draft includes some errors in reporting results. For example, the results of Windham et al. (1991; see page 4-120, lines 21-22) are reported to be adjusted for age, race, education, prior fetal loss, smoking, and number of hours worked, implying multivariable adjustment, whereas data were adjusted for these variables one at a time (see Windham et al. [1991], page 247, paragraph 3).

Finally, the discrepancy in emphasizing spontaneous abortion as the outcome with the strongest evidence of an association with tetrachloroethylene exposure in Chapter 4 and intrauterine growth retardation in Chapter 6 suggests that the evidence on reproductive outcomes was not carefully evaluated.

FINDINGS AND RECOMMENDATIONS

EPA's identification of the key animal and epidemiologic reproductive and developmental studies of tetrachloroethylene appears to be complete, but the committee recommends some reorganization and reconsideration of data to provide a more transparent and balanced characterization of the data. The committee agrees with the selection of the Tinston (1994) two-generation reproductive-toxicity study and the Carney et al. (2006) developmental-toxicity study as supportive of a point of departure and an RfV. EPA's derivation of a comparative RfV based on the developmental toxicity of tetrachloroethylene is an important contribution to the tetrachloroethylene database. However, the committee recommends that EPA revise the chapter to address the specific deficiencies discussed above regarding information presented on the animal reproductive and developmental studies. In particular, the revision should include: (1) a critical analysis of the described studies, including an assessment of the relationship of maternal toxicity to developmental toxicity and the strengths, limitations, and consistency of the various study results; (2) characterization of maternal toxicity (e.g., mild or severe) associated with the studies listed in Table 4-10 and use of consistent nomenclature (ppm or mg/m^3) for listing tetrachloroethylene concentrations; (3) the scientific basis for selecting the Tinston (1994) and Carney et al. (2006) studies as supportive of an RfV; (4) the scientific rationale for selecting the Tinston (1994) study instead of the Carney et al. (2006) study for derivation of the comparative RfV; (5) information on the mode of action for tetrachloroethylene-induced developmental toxicity which addresses the apparent contradictions raised in the committee's review that TCA may be the causative agent; and (6) characterization of the evidence for tetrachloroethylene-induced reproductive and developmental toxicity in animals based on EPA risk assessment guidelines. Stating explicitly whether the animal evidence is sufficient or insufficient for these important end points will help risk managers and others to more readily identify and protect against potential adverse health effects. It will also help to identify data gaps in the tetrachloroethylene database. In addition to revising the chapter, the committee also recommends that EPA consider conducting a bench-mark dose analysis and deriving an RfV based on the Carney et al. (2006) study in addition to, or instead of, the Tinston (1994) study. This will address the potential confounding effects of maternal toxicity at the 1,000 ppm exposure level observed in the Tinston (1994) study.

5

Genotoxicity

Whether tetrachloroethylene and its metabolites are genotoxic (and if so at what doses) is an important consideration in evaluating potential modes of action for carcinogenic effects in the Environmental Protection Agency (EPA) draft Integrated Risk Information System (IRIS) assessment. The evidence on the genotoxicity of tetrachloroethylene is summarized in Section 4.3 of the draft assessment (EPA 2008). The committee found that the publications cited and discussed by EPA are relevant but that the summary does not reflect the entire knowledge base available on the topic and does not provide transparent means for assessing the genotoxicity of tetrachloroethylene itself or its metabolites. The draft IRIS assessment predominantly reports positive studies, whereas good studies that had negative results are not mentioned or in some cases are incorrectly described as having had positive results. The committee therefore recommends that a more balanced, transparent, and inclusive approach be used to consider the evidence. The sections below offer some specific guidance.

ORGANIZATION AND EVALUATION OF DATA

The draft IRIS assessment's consideration of genotoxicity lacks cohesive structure, and the organization of the data presentation should be revised. Specifically, the section should be subdivided into sections on tetrachloroethylene itself, its metabolites, and evidence of indirect genotoxicity. Each section should include a table that lists all primary publications, the results related to tetrachloroethylene in the assays that it was tested in, and comments regarding strengths or weaknesses of each dataset. How the studies were selected should be articulated. It would be helpful if the studies were organized according to the general test systems used; for example, data on nonmammalian systems, in vitro mammalian cells, intact animals, and humans should be delineated separately. A good example of such table may be found in recent monographs of the International Agency for Research on Cancer (IARC). The text that accompanies each table should provide an assessment of the quality of each study cited. At the end of each section, an evaluation of the strength of the evidence of genotoxicity of a particular compound should be included by way of summarizing the totality of data available. Finally, there should be an integrative assessment, including species-specific kinetics and metabolism of tetrachloroethylene and of genotoxicity and mutagenicity in intact animals and humans.

STUDIES OF TETRACHLOROETHYLENE

Nonmammalian Systems

A considerable number of mutagenicity studies of pure tetrachloroethylene that used *Salmonella* strains, *Escherichia coli*, and *Saccharomyces* have been performed with and without exogenous metabolic activation by liver S9 fractions from rats, mice, and hamsters (including animals pretreated with Aroclor or phenobarbital). The results have been essentially negative. The studies should be documented in a table (see above for specific format suggestions). However, when tetrachloroethylene was incubated with purified glutathione *S*-transferase (GST), glutathione, and rat kidney fractions, formation of *S*-(1,2,2-trichlorovinyl) glutathione (TCVG) was found, and mutagenic activity in *Salmonella* was clearly demonstrated as correctly described in the EPA draft.

The committee recommends that EPA also consider the negative results in the National Toxicology Program study (NTP 1986) of sex-linked lethal mutations in *Drosophila*.

Mammalian Cells in Vitro

EPA should describe the mutation study with mouse lymphoma L5178Y cells (NTP 1986), which appears to be the only available mammalian mutation test performed with tetrachloroethylene. This well-done study revealed that tetrachloroethylene at a variety of concentrations, with and without S9 for metabolic activation (but not with GST and rat kidney fractions), did not enhance the frequency of mutations at the thymidine kinase locus. Likewise, investigations of chromosomal aberrations and sister-chromatid exchanges (SCEs) in Chinese hamster ovary (CHO) cells (NTP 1986; Galloway et al. 1987) showed no evidence of tetrachloroethylene-induced genetic activity, although for technical reasons the weight of these studies was somewhat limited. In addition, the negative studies of chromosome aberrations in Chinese hamster lung cells by Sofuni et al. (1985) should be reported.

The work of Hartmann and Speit (1995) is addressed in the draft IRIS assessment, but it is incorrectly quoted in a statement that tetrachloroethylene induced genetic damage, which was not shown. Hartmann and Speit investigated SCEs and DNA integrity (by using the single-cell gel electrophoresis or comet assay) in human blood cells exposed to tetrachloroethylene in vitro. The study was well performed, with negative and positive controls, without and with metabolic activation, and with assay repeats. Although the highest concentration of tetrachloroethylene used in the comet assay was cytotoxic, there clearly was no evidence that tetrachloroethylene at any dose caused increases in SCEs or comet. EPA's review of the study should be corrected.

Concerning the study of Doherty et al. (1996), the EPA draft correctly reports that tetrachloroethylene induced micronuclei in two novel cell lines of human lymphoblastoma origin (h2E1 and MCL-5) through either clastogenic or aneugenic mechanisms. Cells were genetically engineered to express human enzymes (CYP2E1 or CYP1A2, 2A6, 3A4, 2E1) and epoxide hydrolase stably. The committee recommends that EPA acknowledge that those cell lines were not validated as test systems and that other compounds tested in the study, such as hexane and toluene, that are generally regarded as nongenotoxic also led to formation of micronuclei—an indication that the new cell lines may be oversensitive and may provide false-positive results. Micronucleus formation in MCL-5 cells by tetrachloroethylene was confirmed by White et al. (2001), and Wang et al. (2001) found increases in micronuclei in CHO-K1 cells, as mentioned in the draft IRIS assessment.

Tetrachloroethylene's effects on unscheduled DNA synthesis were studied in human fibroblasts (WI-38) (Beliles et al. 1980), in primary hepatocytes from rats and mice (Shimada et al. 1985; Costa and Ivanetich 1984; Milman et al. 1988), and in human lymphocytes (Perocco et al. 1983); the results were mostly negative. Although those studies are limited in performance or reporting, EPA should discuss them to provide a full account of the existing database.

In Vivo Studies in Animals

EPA correctly reports that the study of Walles (1986) showed occurrence of DNA single-strand breaks in liver and kidneys but not lungs of mice 1 hour after intraperitoneal injection of tetrachloroethylene at 650-1,300 mg/kg dissolved in 0.05 mL of Tween 80. EPA fails to mention the full reversibility of that effect at 24 hours. Furthermore, the relevance of the unphysiologic mode of application (intraperitoneal injection in Tween) should be discussed. Tetrachloroethylene is a known irritant of skin and mucosa, and intraperitoneal injection may trigger the release of inflammatory mediators that will stimulate secretion of reactive oxygen species and cytokines in liver and kidney. In addition, the high toxic dose of tetrachloroethylene may produce cell death associated with endonucleolytic DNA fragmentation (Storer et al. 1996). No increase in renal single-strand breaks in DNA was seen 24 hours after oral administration of tetrachloroethylene in rats, but single-strand breaks were enhanced after application of the genotoxins dimethylnitrosamine and diethylnitrosamine (Potter et al 1996).

The EPA draft quotes the paper by Mazullo et al. (1987), which reports low levels of DNA binding 22 hours after intraperitoneal injection of radioactively labeled tetrachloroethylene in mice or rats. Binding was calculated at 2.9

pmol/mg for mouse liver DNA and 0.2-0.5 pmol/mg for rat liver and rat and mouse kidney, lung, and stomach DNA. Thus, there was no evidence of increased binding to rat kidney DNA as misleadingly reported by EPA. Moreover, EPA fails to mention that RNA and protein were labeled much more highly than DNA (up to 420 pmol/mg in the case of RNA). That seriously limits the weight of the study because DNA may have been contaminated by RNA or protein (apparently, DNA was not purified to constant specific activity) and ^{14}C may have been incorporated into DNA via the intermediary metabolism. Overall, those limitations should be taken into account by EPA in the evaluation of the study.

The in vivo micronucleus study in mice by Murakami and Horikawa (1995) is potentially of key importance in the evaluation of tetrachloroethylene's effects on intact organisms. The authors investigated the appearance of micronucleated cells in peripheral blood and liver. However, the draft IRIS assessment is partially incorrect: it reports increased frequencies of micronuclei in peripheral blood reticulocytes after intraperitoneal injection of tetrachloroethylene, but the paper says the opposite (that is, there was no increase in micronuclei in reticulocytes). EPA correctly quotes from the paper in saying that hepatocytes showed small increases in micronuclei when mice received intraperitoneal injections of tetrachloroethylene at high doses 24 hours after partial hepatectomy but not when tetrachloroethylene was injected before partial hepatectomy. The frequency of micronuclei increased less than two-fold but was statistically significant; the positive control diethylnitrosamine produced a 10-fold increase. Several restrictions should be considered by EPA in interpreting the study. The effects were observed at high doses (1,000 and 2,000 mg/kg were effective, but not 500 mg/kg). Given that hepatic toxicity in mice increases from a lowest observed-adverse-effect level of 100 mg/kg (EPA 2008, Section 4.4.2.1), the high doses necessary to enhance micronucleus formation must have been severely toxic to the residual hepatocytes and to the whole organism. The toxic load on the residual liver would have been aggravated by the intraperitoneal tetrachloroethylene application and by the likely release of cytokines and reactive oxygen species. Overall, the small observed increase in micronuclei in mouse hepatocytes might have been due to nonspecific toxic effects. In conclusion, this in vivo study clearly found no increase in reticulocyte micronuclei, and the data suggesting formation of micronuclei in hepatocytes are not convincing.

EPA should mention the in vivo unscheduled DNA synthesis test performed on kidney. Tetrachloroethylene was administered to rats orally (1 g/kg at 0 and 12 hours); at 24 hours, no evidence of unscheduled DNA synthesis in isolated renal cells was observed (Goldsworthy et al. 1988, abstract).

A recent paper by Cederberg et al. (2009) describes the results of an in vivo study in which the alkaline Comet assay was performed on the liver and kidney of CD1 mice treated orally with tetrachloroethylene at 1,000 or 2,000 mg/kg dissolved in corn oil. A slight increase in DNA damage was reported; the effect was significant for one of two end points (tail intensity, but not tail moment) in the liver. No increases were found in the kidney. The study had been performed by a contract laboratory, and the study director had concluded from the same data that tetrachloroethylene did not increase DNA damage because of the inconsistent effects on the two end points, the low magnitude of increases, the high inter-animal variation, and lack of statistically significant increases in a statistical test (Dunnet). Overall, the paper by Cederberg et al. does not present convincing evidence for a genotoxic activity of tetrachloroethylene.

It would also be useful to add the results of studies of hepatic-tumor initiation by tetrachloroethylene although this end point does not necessarily reflect mutagenic activity. When 10 male Osborne Mendel rats were given tetrachloroethylene at 1,000 mg/kg and then phenobarbital as a promoting treatment for 7 weeks (an initiation protocol), the tetrachloroethylene did not induce an increase in the number of gamma-glutamyl transpeptidase-positive cell foci in the liver (Milman et al 1988). Likewise, tetrachloroethylene did not produce liver foci in neonatal female Wistar rats exposed at 2,000 ppm 8 hours/day 5 days/week for 10 weeks (Bolt et al. 1982). Thus, two independent studies did not indicate an initiation potential of tetrachloroethylene in rat liver.

Studies in Humans

Toraason et al. (2003) studied oxidative damage (measured as 8-hydroxydeoxyguanosine [8-OHdG]) in leukocyte DNA of 18 female dry cleaners exposed to tetrachloroethylene and compared it with oxidative damage in 20 female laundry workers who were not exposed to tetrachloroethylene. Blood concentrations in the exposed workers were greater than in unexposed workers by two orders of magnitude. There was a statistically significant reduction in 8-OHdG in the exposed workers and no difference in urinary 8-OHdG or in a urinary lipid peroxidation biomarkers between the two groups. The data from this small sample provide no evidence of oxidative DNA damage under the conditions of the study.

EPA should report the studies by Ikeda et al. (1980a,b), who investigated chromosomal aberrations, SCEs, and modified cell-cycle kinetics in human lymphocytes after 3 days in culture with phytohemagglutinin. Lymphocytes were obtained from 10 workers who had been exposed to tetrachloroethylene and from 11 control subjects. Although no significant effects were found in the exposed group with respect to any of the end points, the limitations of the studies, such as small samples, will need to be considered in evaluating the results.

STUDIES OF METABOLITES OF TETRACHLOROETHYLENE

EPA briefly describes studies that identify TCVG, *S*-(1,2,2-trichlorovinyl)-L-cysteine (TCVC), and *N*-acetyl-*S*-(1,2,2-trichlorovinyl)-L-cysteine (*N*-Ac-TCVC) as bacterial mutagens that act either directly or after activation by rat renal microsomes. It also mentions the induction of unscheduled DNA synthesis by TCVC in a porcine renal-cell line and the key role of renal ß-lyase in the final activation step as demonstrated in these studies.

The EPA draft mentions the positive test for bacterial mutagenicity of tetrachloroethylene epoxide. A discussion of existing studies of the genotoxicity of trichloroacetyl chloride should be added. As to trichloroacetic acid (TCA), the draft states (EPA 2008, p. 4-5) that "as reviewed by Moore and Harrington-Brock (2000), the oxidative metabolite TCA, the major urinary excretion product, exhibits little, if any, genotoxic activity." That statement is followed by brief descriptions of numerous studies of single-strand breaks, which had inconsistent results. Increases in single-strand breaks might have been caused by cytotoxic effects and necrosis at high doses of TCA because of endonucleolytic degradation of DNA (Storer et al. [1996], as reported by EPA). The purpose of the description of studies devoted exclusively to DNA single-strand breaks after exposure to TCA is not clear. The committee recommends integration of the data on single-strand breaks into a balanced review of all available genotoxicity studies of TCA (including a table and a discussion of the studies' strengths and weaknesses) to support the conclusion that TCA exhibits little if any evidence of genotoxicity by an evaluation of the weight of evidence.

Clarity regarding the genotoxicity studies of chloral hydrate and dichloroacetic acid (DCA) is also needed. As recommended earlier, this would be facilitated by an overview of all published data displayed in tables, and there should be a weight-of-evidence evaluation to support EPA's conclusion that chloral hydrate and DCA are genotoxic. That conclusion generally agrees with a recent IARC assessment, but according to IARC (2004), genotoxicity of DCA was limited to high doses that probably are not relevant to tetrachloroethylene carcinogenicity; EPA should consider this argument.

TCVC sulfoxide, another reactive metabolite of tetrachloroethylene, which is nephrotoxic (Elfarra and Krause 2007), does not appear to have been studied for genotoxicity.

EVIDENCE OF INDIRECT GENOTOXICITY

Two studies by Toraason et al. (1999, 2003) are briefly described in the draft IRIS assessment. They revealed no evidence of oxidative DNA damage in rats after a single intraperitoneal dose of tetrachloroethylene at up to 1,000 mg/kg in rats or in humans after occupational exposure to tetrachloroethylene. EPA should add the important information from the animal study by Toraason et al. (1999) that the similar chemical trichloroethylene applied at the same doses as tetrachloroethylene increased oxidative DNA damage in rat liver, whereas tetrachloroethylene did not.

As reported in the IARC (2004) monograph on TCA, the frequency of 8-hydroxydeoxyguanosine-DNA adducts in the liver of B6C3F$_1$ mice was not modified after application of TCA via drinking water (Parrish et al. 1996), was slightly increased after administration through gavage (Austin et al. 1996), and was clearly increased after intraperitoneal injection (Von Tungeln et al. 2002). That comparison of study results again suggests that the route of application (oral vs intraperitoneal) should be considered in evaluating genotoxic effects of tetrachloroethylene and its metabolites.

FORMATION OF REACTIVE METABOLITES IN ANIMALS AND HUMANS

As described in Section 3 of the draft IRIS assessment, the metabolic flux of tetrachloroethylene through glutathione conjugation and β-lyase cleavage is much lower in humans than in rats. TCVG formation in liver, β-lyase activity in kidney, and *N*-Ac-TCVC excretion in urine are all much lower in humans than in rats (Dekant et al. 1986b; Green et al. 1990; Volkel et al 1998). Furthermore, Pahler et al. (1998, 1999) generated monospecific antibodies to the protein adducts of the reactive intermediates either of the glutathione (GSH) conjugation or the oxidative pathway, namely to *N*-dichloroacetyl-L-lysine and *N*-trichloroacetyl-L-lysine. The antibodies allow determination of the amounts of reactive metabolites formed in the two main pathways. Comparing binding in rat kidney and rat liver subfractions, the dichloro adduct (indicating the GSH conjugation pathway) predominates in the kidney with only faint bands in liver; the trichloro adduct (indicating the oxidative pathway) predominates in the liver. Pahler et al. (1999) also compared protein adducts in rat plasma and human plasma obtained from six volunteers. Both adducts were present in rat plasma; in human plasma, the dichloro adducts were below the detection limit, and the trichloro adduct was much lower than in rat plasma. It can be calculated from the data that after exposure to tetrachloroethylene at the same concentration (40 ppm) and duration (6 hours), dichloro adducts were at least 40-fold lower in human plasma than in rat plasma. Trichloro adducts were not quantifiable with gas chromatography for technical reasons (Pahler et al. 1999).

Overall, those results show that humans produce smaller amounts of the reactive metabolites; this is consistent with the overall greater metabolism of tetrachloroethylene in rats. A possible risk of mutagenic effects posed by tetrachloroethylene metabolites with known genotoxic activity should therefore be substantially lower in humans than in rats. However, not all possible metabolites have been assessed for mutagenic activity, and techniques for identifying some metabolites in human samples are not readily available.

Generally, the committee recommends that EPA integrate the qualitative and quantitative data from toxicokinetic, metabolic, and toxicodynamic studies in its assessment of the current knowledge of the toxic potential of tetrachloroethylene and specifically in its mode-of-action considerations.

CELL-TRANSFORMATION ASSAYS

The committee recommends that EPA include at least the more recent cell-transformation studies of tetrachloroethylene (Tu et al 1985, Milman et al. 1988).

FINDINGS AND RECOMMENDATIONS

In vitro studies did not provide evidence of mutagenic activity of tetrachloroethylene in mouse lymphoma cells or in bacterial and yeast mutation assays except in the few tests in which metabolites of the GSH pathway were generated, and no increases in chromosomal aberrations and SCEs were found in CHO cells. Tetrachloroethylene did not increase SCE and comet formation in human blood cells (this was incorrectly reported in the EPA draft); increases in the frequency of micronuclei were found in genetically altered human lymphoid cell lines and in a CHO cell line. In vitro studies of unscheduled DNA synthesis were mostly negative.

The key question is whether the reactive metabolites of tetrachloroethylene are formed and become available to sensitive cells in vivo and have genotoxic effects in intact organisms. Tetrachloroethylene did not induce unscheduled DNA synthesis in rat kidney. It induced single-strand breaks in mouse liver and kidney at 1 hour but not at 24 hours after intraperitoneal injection and not in rat kidney 1 day after oral administration. The increase at 1 hour may be nonspecific because of intraperitoneal application and high doses. Tetrachloroethylene did not increase micronucleated reticulocytes in peripheral blood of mice (this was incorrectly reported in the EPA draft) and did not increase micronucleated hepatocytes when administered before partial hepatectomy. When injected after partial hepatectomy, tetrachloroethylene slightly increased micronucleus formation, but this effect may be nonspecific because of severe liver toxicity caused by the high doses of tetrachloroethylene and the intraperitoneal application of this irritant substance. A study with ^{14}C-labeled tetrachloroethylene suggested a low level of binding to mouse liver DNA and even less to rat liver DNA and mouse and rat kidney, lung, and stomach DNA. These effects are considered nonspecific because DNA was not purified to constant radioactivity and because labeling via the intermediary metabolism appeared likely. In humans exposed to tetrachloroethylene, no evidence of genetic alterations was noted, although the studies are of limited weight. Two studies in rats found no evidence of tumor-initiating activity of tetrachloroethylene (when liver foci were used as the end point).

In conclusion, there is no convincing evidence that tetrachloroethylene has important genotoxic or mutagenic activity in intact organisms. The committee agrees with EPA's conclusion that several metabolites of tetrachloroethylene are clearly genotoxic: TCVG, TCVC, *N*-Ac-TCVC, tetrachloroethylene oxide, DCA, and chloral hydrate. However, it is still questionable whether the metabolites of tetrachloroethylene play an important role in the mode of action of tetrachloroethylene carcinogenesis (see Chapters 6-8) in view of the absence of convincing evidence of mutagenic and tumor-initiating activity of tetrachloroethylene in vivo. Additional studies of genotoxicity in vivo with state-of-the-art methods would be valuable.

As noted above, the committee recommends that EPA provide an expanded and more integrated discussion of the genotoxicity data. The presentation could be improved by the use of tables detailing the primary evidence, by separate discussion of the genotoxic evidence on tetrachloroethylene and its metabolites, and by a more critical analysis of the studies.

6

Hepatic Toxicity and Cancer

This chapter reviews information presented in the Environmental Protection Agency (EPA) draft Integrated Risk Information System (IRIS) assessment of the toxic and carcinogenic effects of tetrachloroethylene on the liver. The metabolism of tetrachloroethylene by the liver is critical for its toxicity and carcinogenicity in that organ. The major metabolites of tetrachloroethylene responsible for hepatic effects are formed by the oxidative metabolic pathway (see Chapter 2 for an overview of toxicokinetics). The following sections address hepatotoxicity and hepatocarcinogenicity separately, but they are not necessarily independent end points. This information is considered in the context of the other evidence on carcinogenicity in Chapter 11, where EPA's assessment of carcinogenic risks posed by tetrachloroethylene is evaluated.

HEPATOTOXICITY

Animal Studies

The draft IRIS document on tetrachloroethylene points out that hepatotoxicity associated with tetrachloroethylene has been shown in rodents in several studies. A number of studies have been conducted with acute administration, but the draft correctly focuses on subchronic and chronic exposures, particularly those involving inhalation as a route of administration. Most of the toxicologic findings focus on increased liver weight, hypertrophy, and histologic lesions, including necrosis.

Damage to the liver by all or most of the chlorinated hydrocarbons has been demonstrated. Tetrachloroethylene is a weaker hepatotoxic agent than, for example, carbon tetrachloride and chloroform; this was shown by studies conducted in the middle 1960s (Klaassen and Plaa 1966, 1967).

The IRIS document overemphasizes a few studies. One is that by Kjellstrand et al. (1984), which is also mentioned in Chapter 3, on neurotoxicity. According to that study, exposure to tetrachloroethylene at 9 ppm for 30 days caused a significant increase in liver weight (not corrected for body weight) in mice. The study also reported an increase in plasma butyrylcholinesterase (BuChE) in mice exposed to tetrachloroethylene at over 9 ppm for 30 days. Although the importance of the change in BuChE is not clear, the exposure in the study was so much lower than those in the other studies cited by EPA that it is important in considering the noncarcinogenic liver end points. EPA does not note that the increase in BuChE at 9 ppm was not significant (it was significant only at 37 ppm and above). It would be valuable for EPA to discuss this study critically in comparison with others in which much higher lowest observed-adverse-effect levels were found. In particular, it should be mentioned that increased BuChE in the Kjellstrand et al. study occurred at 37 ppm only when the exposure was continuous for the entire period, not when exposure at this concentration was intermittent, whereas other studies have involved intermittent exposure (usually 3-6 hours/day). Therefore, the total dose per mouse in the Kjellstrand et al. study must have been several times higher than that in other studies, and the information given in the draft (p. 4-12 and Table 4-2 on p. 4-14) is misleading. It would also be useful for EPA to discuss the quality of studies (for example, deficiencies in reporting by Kjellstrand et al.) and the toxicologic meaning, if any, of the reported effects. Furthermore, the increase in BuChE as a toxic effect does not appear to have been considered important by other investigators, on the basis of citations of the Kjellstrand et al. paper, nor does the effect seem to have been reported by others. Thus, a more critical analysis of the study is necessary to determine the significance of its findings in comparison with other reports of hepatotoxicity that required higher exposure concentrations.

The National Toxicology Program (NTP 1986) and Japan Industrial Safety Association (JISA 1993) studies lend some support to the possibility of hepatotoxicity associated with exposure to tetrachloroethylene. In the NTP 13-week study in rats, hepatotoxicity was evidenced as congestion in the liver. In the 13-week study in mice, there was leukocytic infiltration, centrolobular necrosis, and bile stasis in animals exposed to tetrachloroethylene at 400, 800, or 1,600 ppm. Liver degeneration was observed to occur in a dose-dependent fashion in the 2-year study in mice. In the JISA study, there was an increase in spongiosis hepatitis in Crj:BDF_1 mice, but it is a common finding in these mice and is likely to be unrelated to chemical exposure. Hyperplasia was not statistically significantly increased; there were increases in angiectasis and central degeneration.

In updating and revising the draft IRIS assessment, EPA should include a new 30-day gavage study in Swiss Webster mice given tetrachloroethylene at 150, 500, and 1,000 mg/kg/day (Philip et al. 2007). The metabolism of tetrachloroethylene and its toxicity were examined. That is one of the few studies that were conducted with oral administration and repeated dosing. The investigators found that hepatic injury peaked at 7 days but then was repaired. That suggests that single-dose studies demonstrating hepatic damage on the basis of measurements made after short periods might not mimic the effects of repeated dosing.

Human Studies

EPA also discusses hepatotoxicity in humans. Most of the studies cited in the IRIS draft involved dry-cleaners and found no evidence of an association. However, the EPA document gives undue weight to a couple of studies. One (Brodkin et al. 1995) used sonographic analysis of scattering of fat in liver. This was the only study to report such effects in tetrachloroethylene exposed populations and the importance of the fat changes as an indicator of toxic response is unclear. Furthermore, serum transaminases were not increased in the exposed population. Thus, interpretation of the result is difficult. EPA also considers the study of Gennari et al. (1992). They reported an increase in gamma-glutamyltransferase-2 in tetrachloroethylene-exposed dry-cleaners. The relevance of that finding as an indicator of hepatotoxicity is unclear. The investigators did not find any other indicators of hepatotoxicity despite an extensive serum-enzyme profile. It is likely that the concentrations of tetrachloroethylene that humans were exposed to in those studies were too low to induce frank hepatotoxicity. Further studies are needed.

HEPATOCARCINOGENICITY

Animal Studies

The NTP (1986) and JISA (1993) studies showed, as is the case with many of the halogenated solvents, that there is a dose-dependent increase in hepatic tumors after exposure to tetrachloroethylene in both sexes of mice but not in rats. The draft IRIS assessment's section on hepatic carcinogenicity is written reasonably well in a descriptive sense, with regard to the style of the presentation of the cancer-relevant results of long-term studies with tetrachloroethylene. However, the presentation would benefit if the table on page 5-37, which now gives cumulative tumor incidence, were expanded to include information on species; strain; dose; duration; incidence and multiplicity of adenomas, carcinomas, and other hepatic tumors (such as hemangiosarcomas); and the literature cited.

Tetrachloroethylene induces hepatocellular carcinomas and adenomas in mice. The yield of tetrachloroethylene-induced hepatocellular carcinomas is statistically significant in both male and female B6C3F$_1$ mice after either oral or inhalation exposure. Both male and female Crj:DBF$_1$ mice also have an increased incidence of hepatocellular carcinomas after inhalation exposure to tetrachloroethylene. The earlier studies of the National Cancer Institute (NCI 1977) were repeated, and the findings were confirmed by Nagano et al. (1998). As discussed in more detail below in the section on mode of action, some metabolites of tetrachloroethylene—including trichloroacetic acid (TCA), dichloroacetic acid (DCA), and chloral hydrate (if it is formed)—cause hepatic cancer in mice, and DCA causes hepatic cancer in rats. In the study by Nagano et al., both males and females incurred dose-related increases in incidences of hepatic carcinoma and combined hepatic adenoma and carcinoma.

A difficulty in interpreting the significance of the mouse hepatic tumors is that they have a high spontaneous background incidence in mice. Such tumors have been commonly encountered after exposure to other halogenated solvents, such as dichloromethane, trichloroethylene, tetrachloroethane, carbon tetrachloride, and 1,1,2-trichloroethane.

The curious observation of hepatic and splenic hemangiosarcomas reported in male mice in one of the tetrachloroethylene mouse bioassays (JISA 1993) is mentioned several times in the EPA draft as a potentially important finding; however, there is little discussion of these tumors, the potential mode of action, or the relevance to human risk. Reference to the tumors is presented in Figure 5-14, Table 5-5, and Table 5-9. The analysis is complicated by the fact that the JISA report does not describe the tumors as hemangiosarcomas, but rather as hemangioendothelioma; this term is usually associated with benign tumors, but JISA lists it as a malignant hepatic tumor in male mice. The term is also used for both benign and malignant tumors of the spleen. Furthermore, because of the cell types involved, the hepatocellular carcinomas being of hepatocellular origin and the hemangiosarcomas being of endothelial-cell origin, it is scientifically inappropriate to lump these tumors in with carcinomas, as is done by EPA (Figure 6.4 and Table 6.4).

Human Studies

Available epidemiologic evidence does not support an association between tetrachloroethylene and hepatic cancer. Two cohort mortality studies of dry-cleaner union members (Ruder et al. 2001; Blair et al. 2003) and a large (N = 77,965) cohort mortality study of aerospace workers (Boice et al. 1999) report no association with hepatic-cancer mortality. A sizable subcohort (N = 2,631) of the aerospace workers routinely exposed to tetrachloroethylene had a standardized mortality ratio of 2.05 (95% confidence interval [CI], 0.83-4.23) on the basis of seven observed deaths. However, an analysis that used an internal cohort referent population to reduce confounding yielded no overall association and no exposure-response relationship. Because hepatic cancer is fatal, assessments of mortality represent the burden of the disease in the population. Essentially null associations are reported in studies of incident cancers in laundry workers residing in Nordic countries. In the one study cited (Lynge et al. 1995) that reported an increased standardized incidence ratio (SIR) for hepatic cancer in women (2.7; 95% CI, 1.5-4.5; 14 observed cases, all cases were in laundry workers, and no cases were observed in dry-cleaning workers, whose exposure to tetrachloroethylene is more likely. (The EPA document does not cite this correctly in Table 4B-1a; the reference should be to Lynge et al. 1995, which is an update of Lynge and Thygesen 1990.) Those studies identified laundry and dry-cleaning workers on the basis of the census in 1970 and 1980, so the extent of exposure is unknown. Several population-based case-control studies of hepatic cancer and exposure to solvents (determined by occupation) have been conducted over the last 30 years. Overall, they have not reported an association between tetrachloroethylene and hepatic cancer. Some evidence is suggestive of an association between solvent exposure and laundry work and hepatic cancer in women, but the exposure models for these studies are crude, and methods of control selection raise questions about the validity of the results.

The draft IRIS assessment does not use that limited evidence of an association between tetrachloroethylene and hepatic cancer as supportive of classifying tetrachloroethylene as a carcinogen. The argument that human epidemiologic evidence supports classification as "likely to be a carcinogen" is limited to other cancers, specifically esophageal and lymphoid cancers. The exclusion of hepatic cancer as supporting evidence is appropriate.

Mode of Action

The draft assessment describes the mode of action (MOA) of tetrachloroethylene's hepatic toxicity and carcinogenicity in several places. The most comprehensive description of the available body of information and identification of potential key events in the MOA are included in Section 4.4.4. The MOA summary is provided in Section 4.10.3, including Table 4-13; Appendix 4A details the EPA-conducted analysis of the consistency between carcinogenicity of tetrachloroethylene and that of one of its major oxidative metabolites, TCA; and Section 6.1.5 includes a short summary of the liver MOA with regard to the human hazard potential of tetrachloroethylene.

EPA concludes that "the MOA for tetrachloroethylene-induced mouse liver cancer is not well understood, and it is highly likely that more than one MOA is operative" (EPA 2008, p. 4-16). In support of that conclusion, EPA describes pathways that could lead to hepatic tumors but does not clearly describe the weight-of-evidence approach for determining the key elements in the tumorigenicity of tetrachloroethylene for the possible MOAs presented. The difficulty in characterizing the MOA is not surprising given the complexity of the metabolic pathways for tetrachloroethylene, the closely related chlorinated solvent trichloroethylene, and their common primary oxidative metabolites, TCA and DCA. The following major events are put forth as plausible components of the MOA of hepatic carcinogenicity of tetrachloroethylene (in no particular order with regard to a temporal sequence):

- Metabolism of tetrachloroethylene to TCA and DCA, which are both considered ultimate hepatotoxic metabolites.
- Activation of peroxisome proliferator-activated receptor-α (PPARα) and the downstream cascade of the molecular events that include induction of peroxisomes, increase in cell proliferation, and decrease in rates of apoptosis.

- Other nongenotoxic events, such as promotion of growth of previously initiated foci, changes in epigenetic status, cytotoxicity, and oxidative stress.
- Genotoxic events, such as DNA damage by tetrachloroethylene metabolites or chromosomal aberrations.

Although the discussion of the PPARα-mediated events and their possible roles in species differences with regard to the hepatocarcinogenic potency of tetrachloroethylene is extensive, other important potential MOAs or key events are largely overlooked. For example, the possible role of epigenetic changes caused by TCA and DCA is mentioned, but there is little discussion of the studies that have been conducted on this subject. Similarly, cytotoxicity and secondary oxidative stress that may result from microsomal enzyme induction are insufficiently considered. Adding such discussions would strengthen EPA's MOA analysis and conclusions.

That TCA is the major urinary metabolite of tetrachloroethylene and is a mouse hepatocarcinogen suggests that the hepatocarcinogenicity of tetrachloroethylene is due in part to TCA. DCA is another tetrachloroethylene urinary metabolite that is formed both in the oxidative pathway by dechlorination of TCA and, in organs other than the liver, in the glutathione (GSH) pathway. DCA is known to cause hepatic cancer in both rats and mice, so it is possible that DCA contributes to the hepatocarcinogenicity of tetrachloroethylene, although it is not certain to what extent it contributes in that little of it is produced and it is produced primarily in the kidney. Early studies that reported finding DCA as a metabolite may have overstated the amount formed because of problems with analytic methods (Ketcha et al. 1996). Later studies showed very small amounts of DCA, if any, being formed from tetrachloroethylene. Chloral hydrate (if it is formed) is a mutagen and is a hepatocarcinogen in mice and might contribute to the hepatocarcinogenicity of tetrachloroethylene. In addition, metabolites formed from the GSH pathway, such as trichlorovinylglutathione, which is further metabolized by β-lyase in the kidneys, are also genotoxic.

The multiplicity of metabolites formed from tetrachloroethylene that are toxic and carcinogenic—TCA, DCA, tetrachloroethylene oxide, trichloroacetyl chloride, and possibly chloral hydrate—makes it difficult to determine the MOA of hepatocarcinogenicity of tetrachloroethylene. Indeed, there may not be enough data to determine quantitatively the extent to which each metabolite contributes to tetrachloroethylene-induced hepatotoxicity. Perhaps a summary of the available information on hepatocarcinogenicity of TCA, DCA, and chloral hydrate administered alone or in combination with other compounds—for example, from studies of Bull et al. (1990, 2002, 2004) on mixtures and coadministration with gadolinium chloride—should be included in the IRIS assessment and in tabular form (e.g., see table in NRC 2006a, pages 149-156) to better assess the data.

Although the consideration of the metabolic activation of tetrachloroethylene and the comparison with TCA-induced carcinogenesis are useful, the dose-response information in the draft on tumor formation after TCA administration (Table 4A-2) suggest that very high concentrations of TCA are needed to cause hepatic tumors—far beyond what would be generated after tetrachloroethylene administration.

The peroxisome-proliferator MOA is discussed in great detail. The key events associated with the known links between peroxisome-proliferator chemicals in general and rodent hepatic cancer are identified, and appropriate literature references are included. However, no data or weight of evidence criteria specifically on tetrachloroethylene are provided, and the lack of coherent flow in the document detracts from the intended message. The document might be improved by organizing the information into sections that make clear (1) what parts of this MOA are based on studies with other model peroxisome proliferators, (2) what data are available to support this MOA for tetrachloroethylene, (3) for TCA, (4) the rationale for species differences, and (5) the relevance of this MOA to mouse hepatic tumors induced by tetrachloroethylene or to human risk.

As presented, the draft IRIS assessment seems to be more concerned with critiquing the current dominant view in the field that the peroxisome-proliferator MOA may not be relevant to human hepatocarcinogenesis than with providing evidence of links between tetrachloroethylene and this MOA. The general criticism of the MOA with regard to its relevance to humans is warranted, although it should be expressed in milder terms, and it points correctly to several historical and recent lines of evidence that suggest important inconsistencies that challenge the paradigm of the central role of PPARα in rodent, but not human, hepatocarcinogenesis. However, as pointed out above, the data linking tetrachloroethylene to this MOA are weak to begin with and come largely from studies of trichloroethylene and TCA, not tetrachloroethylene itself. The idea that there are deficiencies in our knowledge of tetrachloroethylene should be made more prominent. Similarly, the discussion of "tetrachloroethylene and PPARα MOA" and the discussion of "relevance of the PPARα MOA to human liver carcinogenesis" should be separated more clearly by EPA.

The discussion of the strain and species differences in the peroxisome-proliferation effect of TCA is rather limited. TCA is capable of inducing peroxisome proliferation in the rat, but tetrachloroethylene does not. In addition, the issues of PPARα transactivation by tetrachloroethylene, related chemicals, and their key metabolites and of species differences are important for the discussion of the MOA. Again, a critical look at the quantitative differences in metabolic activation of tetrachloroethylene to TCA between mouse and rat, species that are generally believed to be almost equally sensitive to peroxisome proliferation and differences between mouse and rat in hepatic cancer induced by other compounds of this class should be provided. Specifically, EPA may consider performing additional analyses with the rat data similar to those with the mouse data in Appendix 4A and including a table showing the quantitative differences in affinity between mouse, rat, and human PPARα of tetrachloroethylene and its key metabolites in comparison with the known peroxisome proliferators. Such analyses and data would greatly facilitate the discussion of quantitative differences between compounds and between species.

The study by Nakajima et al. (2000) is only mentioned in passing on page 4-26 of the draft assessment. It should be discussed in greater detail, especially the data on sex differences and mechanistic considerations. It provides a possible mechanistic explanation for sex differences in susceptibility to carcinogenesis by tetrachloroethylene—information that is important for the discussion of the complexities of and uncertainties in the MOA.

The dose-response relationship in Section 4.4.4.3.6 touches on the important issue. However, the arguments are not supported by adequate literature citations, and the only paper cited is a broad review article, not a primary source of the data. Section 5 contains ample information on dose-response relationships, so appropriate cross-referencing should be included in Section 4.4.4.3.6.

The discussion on nonliver targets in humans that may involve PPARα MOA is interesting, but it is too brief and is not adequately linked to the rest of the chapter to have an appropriate impact. The arguments presented in Section 4.4.4.3.8 may be substantiated by providing a quantitative comparison of PPARα transactivation potential by tetrachloroethylene and its metabolites, as suggested above. Similarly, the discussion of the potential role of PPARγ is inadequate. Specifically, it should be noted that PPARγ may be an important gene for human hepatocellular carcinogenesis.

The committee agrees with EPA that the MOA of tetrachloroethylene-induced hepatic tumors is not clear. Many toxic metabolites are formed from tetrachloroethylene. Hence, it is likely that key events from several pathways operate in tetrachloroethylene-induced hepatocarcinogenesis. It is likely that TCA, DCA, and chloral hydrate (if it is formed)—which are carcinogens in rodents—contribute to tetrachloroethylene-induced hepatocarcinogenesis. It is also likely that mutagenic metabolites of tetrachloroethylene formed via the cytochrome P450 and GSH pathways (tetrachloroethylene-epoxide, TCA, DCA, and TCVG) contribute to hepatocarcinogenesis. And it is possible that activation of PPARα and consequent peroxisomal proliferation; genotoxic events induced by tetrachloroethylene metabolites, including chromosomal aberrations; and other nongenotoxic events—such as promotion of growth of previously initiated foci, changes in epigenetic status, and oxidative stress—may all contribute to the overall MOA through several simultaneous mechanisms. The hypothesis that the mutagenic metabolites of tetrachloroethylene (tetrachloroethylene-epoxide, TCA, DCA, chloral hydrate [if it is formed], and TCVG) initiate carcinogenesis and that tetrachloroethylene-induced promotion of initiated foci, cytotoxicity, and epigenetic events promote carcinogenesis cannot be ruled out.

SUMMARY

As with other halogenated solvents, there is evidence in a number of species that tetrachloroethylene can cause liver damage. This was well described by

EPA in the drat IRIS assessment. Two rodent bioassays have demonstrated that high doses of tetrachloroethylene produced liver tumors in mice. While there is clear evidence that this occurs, the basis for their occurrence is not clear and may actually involve more than one MOA. This makes the determination of the relevance to humans more difficult. This is particularly true with respect to the importance of PPARα as the predominant or sole MOA, which led to a split opinion among committee members and a dissenting statement (see Appendix B).

Further studies are needed to define the MOAs for tetrachloroethylene-induced liver tumors, with particular emphasis on the importance of PPARα and whether species difference might exist. In addition, further study is needed to determine the relative roles of metabolites of tetrachloroethylene in tumor development. This may require the development of better analytical methods to detect some metabolites.

7

Renal Toxicity and Cancer

This chapter reviews information presented in the Environmental Protection Agency (EPA) draft Integrated Risk Information System (IRIS) assessment of the toxic and carcinogenic effects of tetrachloroethylene on the kidney. The metabolism of tetrachloroethylene by the kidney is critical for its toxicity and carcinogenicity in that organ. The major metabolites of tetrachloroethylene responsible for renal effects are formed by the glutathione metabolic pathway (see Chapter 2 for an overview of toxicokinetics). The following sections address renal toxicity and carcinogenicity separately, but they are not necessarily independent end points. This information is considered in the context of the other evidence on carcinogenicity in Chapter 11, where EPA's assessment of carcinogenic risks posed by tetrachloroethylene is evaluated.

HUMAN STUDIES

Renal Toxicity

The draft IRIS assessment notes that published information on renal toxicity in humans is not well developed. That is because typical screening tests that use plasma are insensitive. For instance, blood urea nitrogen and creatinine, which accumulate in plasma when glomerular filtration is diminished, do not increase until renal function is about half of normal, and urinalysis is not typically performed. Epidemiologic studies of the effects of tetrachloroethylene exposure on renal function have been reported, and EPA summarizes the findings in a table. The discussion focuses on urinary proteins that are indicative of tubular damage, because β-lyase is found in the proximal tubule. The strengths and weaknesses of the various studies are noted by EPA, and consistencies and inconsistencies are discussed. In general, different reports examined different urinary proteins, which have different sensitivity and selectivity as markers of tubular function. Estimated exposure differed among the reports, as did the number of subjects. Effects on glomerular function, as assessed on the basis of albuminuria, are discussed briefly. The draft IRIS assessment notes that the results are contradictory. It should also note that some albumin is normally filtered, so small increases in the amount of albumin in the urine can be indicative of tubular damage (the result of failure to reabsorb the small amount filtered). EPA's table should also include the negative findings on albumin in studies by Verplanke et al. (1999) and Lauwerys et al. (1983) and on total protein by Vyskocil et al. (1990). EPA concluded that the epidemiologic studies provided evidence suggestive of subtle damage in renal tubules. The committee agrees with that assessment.

Renal Carcinogenicity

Several types of epidemiologic studies have been used to explore a possible association between jobs in which workers are exposed to tetrachloroethylene and renal-cell carcinoma (RCC), including cohort mortality studies, case-control studies, and nested case-control studies. Ultimately, the methodologic challenges of studying such a rare cancer as RCC, assessing tetrachloroethylene exposure accurately, and evaluating inconsistencies in results among studies limit the conclusions that can be drawn from the epidemiologic data. Most of the studies either did not have explicit information about exposures or had considerable methodologic limitations.

Pesch et al. (2000) conducted a population-based case-control study in Germany that estimated tetrachloroethylene exposure with a job-exposure matrix (JEM) and a job-task exposure matrix (JTEM). The latter is usually superior for estimating specific exposures. The data were acquired in in-person interviews, so information on occupational history was obtained and confounding covariates (such as smoking) were well measured. An increased odds ratio (OR) for tetrachloroethylene exposure was observed in men who had a medium exposure index (OR, 1.4; 95% confidence interval [CI], 1.1-1.7) or a substantial exposure index (OR, 1.4; 95% CI, 1.1-2.0) on the basis of the JEM. However, the results based on the JTEM were less convincing (OR, 1.2; 95% CI, 0.9-1.7 and OR, 1.3; 95% CI, 0.7-2.3 for medium and substantial exposure, respectively). In contrast, no association was observed in women on the basis of the JEM, but a positive albeit imprecise association was observed on the basis of the JTEM for medium and substantial exposure (OR, 2.2; 95% CI, 0.9-5.2 and OR, 2.0; 95% CI, 0.5-7.8, respectively). Those variable results are representative of inconsistencies among studies. Lynge et al. (2006) (listed in Table 4B-4 of the EPA draft but not discussed in the renal-cancer section) conducted a nested case-control study in four Scandinavian countries in a cohort of about 47,000 persons employed in the laundry and dry-cleaning industry as of 1970 and followed through 1997-2001 to identify incident cancers. Multiple cancers were assessed, including 56 RCC cases in men and 154 in women. The cohort was divided into those who were not exposed to the dry-cleaning process, dry-cleaners and other exposed workers, and others working in dry-cleaning. Risk was also estimated by duration of employment in dry-cleaning occupations. Tetrachloroethylene was the most commonly used solvent in dry-cleaning during the study interval; the mean concentration over the interval of the study was estimated as 24 ppm. The adjusted relative risk of RCC for dry-cleaners compared with unexposed workers was 0.67 (95% CI, 0.43-1.05) on the basis of 29 cases in the exposed. There was no evidence of increasing risk with increasing duration of employment as dry-cleaners. Mandel et al. (1995) pooled data from a multicenter international case-control study of RCC; the study was conducted in six centers in five countries (Australia, Denmark, Germany, Sweden, and the United States) and included 1,732 cases and 2,309 controls. Occupational histories, collected in in-person interviews, were used to estimate exposures to specific chemicals or tasks. The study reported an increased OR of 1.4 (95% CI, 1.1-1.7) associated with exposure to dry-cleaning solvents, but no exposure response was observed on the basis of duration of exposure.

Several other studies, although methodologically sound, were too small or did not have sufficient information about exposure to be informative (Aschengrau et al. 1993; Mellemgaard, et al. 1994; Schlehofer et al. 1995; Dosemeci et al. 1999).

There are inconsistencies in the draft IRIS assessment. Nine studies are listed as larger case-control studies. Of them, EPA judged the case-control studies of Aschengrau et al. (1993), Partanen et al. (1991), and Pesch et al. (2000) to be of high quality, citing exposure information, adequate control of confounding, and histologic confirmation. It is then noted that "these two case-control studies carry greater weight than observations in the other case-control studies identified in Table 4B-4." The Aschengrau et al. study is not listed in Table 4B-4; and this suggests that the Partanen et al. and Pesch et al. studies are those considered to be the studies given greater weight. The point should be clarified. The Lynge et al. (2006) study is not discussed in the "Kidney Cancer in Humans" section of the draft IRIS assessment.

Overall, the epidemiologic literature provides little support for a causal association between tetrachloroethylene exposure and cancer of the kidney. Study results are inconsistent. In addition, those studies that tried to assess dose-response by using the imperfect surrogate of "duration of exposure," found no association between duration and risk. EPA's assessment of the data appropriately labels the evidence supporting an association between tetrachloroethylene and renal cancer as "limited," and the epidemiologic evidence does not appear to weigh heavily toward classifying tetrachloroethylene as a likely carcinogen.

ANIMAL STUDIES

Renal Toxicity

The draft IRIS assessment summarizes the studies of tetrachloroethylene toxicity across species, sexes, and routes and durations of exposure. Significant renal toxicity has been observed in lifetime bioassays in rats and mice of both sexes (NCI 1977; NTP 1986; JISA 1993). Degenerative changes in the proximal tubule are reported as cloudy swelling, fatty degeneration, and necrosis of the epithelium. Some tubules were filled with hyaline casts; inflammatory cells, fibrosis, and focal mineralization were also reported. Effects of shorter exposures depended on route, duration, and dose. In short term (28-42 days) gavage studies, male rats showed signs of renal damage (Green et al. 1990; Philip et al. 2007). Inhalation exposure of male and female rats and mice to tetrachloroethylene for 28 days caused no effects at 400 ppm, and exposure of male rats for 10 days at 1,000 ppm resulted in an increase in hyaline droplets (Green et al. 1990). Inhalation exposure to tetrachloroethylene for 13 weeks resulted in karyomegaly in male and female mice but not in rats (NTP 1986); the response was minimal at 200 ppm and increased in severity with exposure concentration.

Renal Carcinogenicity

Renal-tubular adenoma and carcinoma were observed in male rats in the NTP (1986) bioassay and to a lesser extent in the Japan Industrial Safety Association (JISA 1993) studies. Tetrachloroethylene caused a low rate of induction of renal tumors in rats; although the yield at the high dose was not statistically significant. In the NTP bioassay, induction of renal tumors was dose-dependent. The incidence was 1 of 49 in the control group, 3 of 49 in the 200-ppm group, and 4 of 50 in the 400-ppm group. There are wide confidence limits on the data, and some of the error bars approach zero. There is a very low spontaneous incidence of renal tumors in Fischer 344 rats (Haseman et al. 1998). Induction of renal tumors in rats by tetrachloroethylene is therefore easily observed against a low background. In addition, the controls had only benign tumors, not malignant tumors, whereas the high-dose group had two malignant tumors. In the draft IRIS assessment, EPA calculates the chance that two animals will have a rare tumor to be less than 0.001, giving biological relevance to the finding. Maltoni and Cotti (1986) observed no increase in kidney tumors following tetrachloroethylene administration by gavage to male Sprague-Dawley rats. Overall, the dose-dependent induction of renal tumors in one experiment against the low background incidence of renal tumors in rats exposed to tetrachloroethylene indicates that tetrachloroethylene can induce renal tumors in rats. After integrating the results of the studies, the committee concluded that tetrachloroethylene induces renal tumors in rats. EPA considers the renal tumors to be suggestive of an effect and notes that it is similar to the effects of other chlorinated ethanes and ethylenes. The committee agrees with EPA's assessment.

Mode of Action

EPA considered key events and potential modes of action for renal-tumor formation following tetrachloroethylene exposure and concluded that the mechanisms are not understood.

The draft IRIS assessment discusses an $\alpha_{2\mu}$-globulin nephropathy mode of action of tetrachloroethylene-induced renal carcinogenesis in detail. Renal tumors that arise solely by $\alpha_{2\mu}$-globulin nephropathy are not considered relevant to human risk assessment, because $\alpha_{2\mu}$-globulin nephropathy is specific to the male rat. Although hyaline droplets that contain $\alpha_{2\mu}$-globulin have been reported after exposure to high concentrations of tetrachloroethylene, the histopathologic findings reported in the inhalation bioassays were not consistent with the $\alpha_{2\mu}$-globulin-mediated mode of action (NTP 1986; JISA 1993). Gavage bioassay (NCI 1977) showed that histopathologic characteristics were more consistent with $\alpha_{2\mu}$-globulin nephropathy. However, in all these bioassays, similar histopathologic findings in the kidney were reported in female rats and male and female mice. These positive responses are not consistent with the male rat specificity of the $\alpha_{2\mu}$-globulin nephropathy mode of action and therefore contradict a role of $\alpha_{2\mu}$-globulin nephropathy in renal tetrachloroethylene tumorigenesis. The committee agrees with EPA's assessment that $\alpha_{2\mu}$-globulin nephropathy is not supported as a mode of action in tetrachloroethylene-induced renal carcinogenesis.

Tetrachloroethylene can stimulate the peroxisome proliferation response, as indicated by cyanide-insensitive palmitoyl CoA oxidation activity, in the kidneys of mice but not rats (Goldsworthy and Popp 1987). Odum et al. (1988) reported similar findings; mouse kidney samples were pooled for assays, so statistical analysis was not conducted on mouse kidneys. The peroxisome proliferation response does not correlate with tumor response and therefore is not consistent with a role of peroxisome proliferation as a mode of action in renal tumorigenesis. EPA notes that activation of peroxisome proliferator-activated receptors has not been established as a mode of renal tumorigenesis. The committee agrees that the data do not support peroxisome proliferation as a mode of action.

The draft IRIS assessment also considers immunotoxicity and immunosuppression as a mode of action of tetrachloroethylene tumorigenesis. In humans, immune-mediated renal damage is most often seen as damage to the glomeruli. The reports of renal damage in humans are based on abnormal protein in the urine; the pattern of proteinuria is indicative of tubular, not glomerular, damage. Thus, the type of renal damage seen is not consistent with an immunotoxic mode of action. The draft IRIS assessment notes that immune-system-mediated effects of organic solvents and the formation of protein adducts are related to autoimmune diseases, not to immunosuppression and therefore inconsistent with immunosuppression as a mode of action.

Tetrachloroethylene causes toxic nephropathy in high doses, and this was observed in the cancer bioassay studies (NCI 1977; NTP 1986; JISA 1993). EPA considered a mode of action in which renal cytotoxicity and subsequent proliferation—as part of the repair process, not associated with $\alpha_{2\mu}$-globulin—result in renal-tubular neoplasia. Renal toxicity has been observed with various metabolites of tetrachloroethylene (Lash et al. 2007; Elfarra and Krause 2007). Each of the three major metabolic pathways of tetrachloroethylene yields metabolites that are cytotoxic (Dekant et al. 1986c, 1988; Vamvakas et al. 1989a,c;

DeMarini et al. 1994; Werner et al. 1996; Volkel and Dekant 1998; Muller et al. 1998a; Dreessen et al. 2003). Chronic nephrotoxicity has been reported in male rats at the termination of all long-term bioassays but also has been observed in chronic bioassays at 2 years in female rats and both sexes of mice, none of which develop tumors. Despite this inconsistency, it is not possible to rule out a role of chronic toxicity in tumor formation.

The draft IRIS assessment concludes that a mutagenic mode of action cannot be ruled out. The committee agrees with this assessment. A mutagenic mode of action is supported by the findings after exposure to the structurally similar trichloroethylene. Some metabolites derived from *S*-(1,2,2-trichlorovinyl) glutathione (TCVG), the glutathione conjugate of tetrachloroethylene, have been shown to be mutagenic in bacterial systems (Vamvakas et al. 1989a,d) or to cause unscheduled DNA synthesis (Vamvakas et al. 1989c). Others react with DNA in vitro (Muller et al. 1998a,b). *S*-(1,2,2-Trichlorovinyl)-L-cysteine (TCVC) causes a greater response than dichlorovinyl cysteine in mutagenicity tests using *Salmonella* (Dekant et al. 1986c) and in renal toxicity (Birner et al. 1997). Tetrachloroethylene has not been shown to be mutagenic with or without activation by S9 in *Salmonella* or in mammalian cells. However, when tetrachloroethylene was activated with purified glutathione *S*-transferase, glutathione, and rat kidney fractions, TCVG was formed, and consequent mutagenic activity in *Salmonella* was clearly demonstrated, as described by EPA. S9 activation of tetrachloroethylene did not induce mutation in cultured mouse lymphoma L5178Y cells.

SUMMARY AND RECOMMENDATIONS

EPA concluded there is limited evidence that tetrachloroethylene causes cancer in humans, and the committee agrees with this assessment. EPA evaluated bioassay studies to provide evidence suggestive of an effect. The committee considers this and the similarity to trichloroethylene to support the conclusion that tetrachloroethylene induces kidney tumors in rodents. While the mode of action of tetrachloroethylene tumorigenesis is not understood, the $\alpha_{2\mu}$-globulin nephropathy and peroxisome proliferator modes of action are not consistent with experimental results. A mutagenic mode of action cannot be ruled out.

Further studies are needed to determine whether tetrachloroethylene and its metabolites formed from TCVG (TCVC, chlorothioketene, and sulfoxide metabolites) are mutagenic in other mammalian cell assays (mutation to 6-thioguanine resistance in cultured V79 Chinese hamster lung fibroblasts or in Chinese hamster ovary cells). It is possible that any of the metabolites of TCVG contribute to the carcinogenicity of tetrachloroethylene in rat kidney, but this needs to be studied. Further data on the sequencing of DNA from tetrachloroethylene-induced renal tumors for mutations of the von Hippel Landau tumor-suppressor gene, other tumor-suppressor genes and oncogenes, and their downstream effectors (for example, p27 that controls cell-cycle progression) are needed to determine whether TCVG and similar tetrachloroethylene metabolites are causing or contributing to the formation of renal tumors. Finally, a robust physiologically based pharmacokinetic model is needed to evaluate differences between humans and rats in their sensitivity to tetrachloroethylene.

8

Hematopoietic Effects

This chapter reviews information presented in the Environmental Protection Agency (EPA) draft Integrated Risk Information System (IRIS) assessment of the effects of tetrachloroethylene on the hematopoietic system, especially the development of mononuclear-cell leukemia (MCL) in rats and lymphomas in humans. The information is considered in the context of the other evidence on carcinogenicity in Chapter 11, where EPA's assessment of carcinogenic risks of tetrachloroethylene is evaluated.

ANIMAL STUDIES

The draft IRIS assessment proposes to use the finding of MCL in male F344 rats as the most sensitive tumor response, supporting its weight-of-evidence classification of tetrachloroethylene as "likely to be carcinogenic to humans" by all routes of exposure. The use of MCL to support that conclusion is based primarily on two studies: those of the National Toxicology Program (NTP 1986) and the Japan Industrial Safety Association (JISA 1993). Both studies reported that chronic inhalation administration of tetrachloroethylene to male and female F344 rats caused "positive trends" in MCL with increasing dose. As the draft IRIS document correctly points out, the scientific reliability of those studies has been questioned in part because of "high spontaneous background incidences, use of special supplemental analysis to aid in data interpretation, and the relevance of MCL in F344/N rats to human hazard" (p 4-159, lines 21-23). The committee similarly questions the use of the tetrachloroethylene exposure bioassays in the F344 rat for cancer risk assessment for those reasons and others discussed below.

In the NTP (1986) study, F344/N male and female rats were exposed chronically to tetrachloroethylene at 200 and 400 ppm. The incidence of MCL in males was 77% in the 200-ppm group and 74% in the 400-ppm group, and in females 60% and 58%, respectively. The background incidences of MCL in the controls were high (56% in males and 36% in females). Such high backgrounds make it difficult to interpret the biological significance of the increase in the incidence of MCL observed in the treatment groups. Indeed, NTP has decided to stop using its F344/N rat colony in its bioassays for reasons that include the high background rate of MCL (King-Herbert and Thayer 2006). A supplemental analysis performed by NTP considered disease progression, latency, and various statistical treatments. The analysis suggested an increase in tumor incidence over controls at both test concentrations despite the high spontaneous tumor incidence in the controls.

The significance of MCL findings in multiple NTP bioassays that used the F344 rat was the subject of a recent reanalysis by Thomas et al. (2007), which EPA should reference in the draft IRIS assessment. They examined the incidence of leukemia in 2-year bioassays that included untreated male and female F344 rats from 1971 to 1998. They found that background tumor incidence increased substantially, from 7.9% to 52.5% in males and from 2.1% to 24.2% in females, over that period. The analysis also found that MCL responses are highly variable and subject to substantial modulation by dietary factors.

Thomas et al. (2007) also evaluated MCL incidence in male and female rats exposed to 500 chemicals. On the basis of 34 NTP studies that yielded evidence of a chemically related increase in the incidence of leukemia, which included the 1986 NTP study of tetrachloroethylene, the authors conducted a reanalysis of dose-response data by comparing results with four statistical methods: Fisher's test for pair-wise comparison of leukemia incidence between a dose group and a control group, the Cochran-Armitage test for incidence trend, logistic regression for incidence, and life tables for survival-adjusted incidence. Tetrachloroethylene was one of five chemicals shown by the authors to produce leukemia in both sexes of rats. They used the rigid Food and Drug Administration (FDA) statistical criteria for testing dose-related cancer incidences ($p < 0.01$ for pairwise comparison; $p < 0.005$ for trend test). The results in male rats in the 1986 NTP study revealed a significant dose-response trend when analyzed with a life table ($p = 0.004$) assuming that MCL is lethal but a nonsignificant trend with logistic regression ($p = 0.097$) assuming the MCL is nonlethal. Pairwise comparisons revealed dose-related incidences ($p = 0.046$) for both dose groups, and the trend test yielded a p value of 0.034; neither met the FDA criteria for statistical significance. The borderline significance of the trend test and nonsignificance of logistic regression for the latter two comparisons could be explained in part by the fact that the incidences did not follow an incrementally increasing relationship with dose. In female rats in the NTP study, use of a life table ($p = 0.053$), logistic regression ($p = 0.012$), a trend test ($p = 0.018$), and Fisher's test ($p = 0.014$ and 0.022, respectively, for two doses) all revealed a borderline significant dose-related incidence. However, there is inconsistency in statistical significance between the sexes and uncertainty about the shape of the dose-response curve, especially in the lower range of the study. The authors recommended the use of life-table analysis for survival-adjusted leukemia incidence, noting that it is "closer to reflecting the true statistical significance of the carcinogenic effect" than logistic-regression treating dose as linear. Life-table analysis (log-rank test) accounts for time-to-event information, is capable of testing nonlinear dose-response relationships of arbitrary shapes, and is therefore more flexible than the Cochran-Armitage trend test. Survival analysis also is more relevant than logistic regression for more lethal tumors such as MCL. Overall, Thomas et al. showed a moderately significant dose-response relationship for tetrachloroethylene, but this finding should be evaluated by EPA with a weight-of-evidence approach suggested in its 2005 *Guidelines for Carcinogen Risk* before conclusions are drawn.

In the 1993 JISA study, F344/DuCrj rats were exposed to tetrachloroethylene at 50, 200, and 600 ppm. The draft IRIS document focuses on the JISA report for cancer dose-response assessment because the study included a 50-ppm exposure concentration, which is one-fourth the lowest exposure concentration in the 1986 NTP study. As in the NTP study, there was a high incidence of MCL in the controls (22% in males and 20% in females). Against that high spontaneous incidence of MCL, the incidence of MCL in male and female rats exposed to tetrachloroethylene at 50, 200, and 600 ppm was 28%, 44%, and 54% and 34%, 32%, and 38%, respectively. Moreover, the historical rate of MCL for the Japanese laboratory is very high. There was no incremental increase in MCL incidence in female rats with increasing dose. In contrast, EPA concluded that male rats displayed a dose-dependent increase in MCL although in the analysis background values were subtracted from the incidences in animals treated with tetrachloroethylene (Figure 5-6 in the draft IRIS assessment), and this may lead to a false impression. Such manipulation of data is not widely accepted in statistical practice, because it artificially reduces the uncertainty caused by the variation in the background rate. As noted in reviews by Caldwell (1999) and Ishmael and Dugard (2006), the unusually high background rate of MCL in control (untreated) rats weakens the ability to separate the background response from possible chemically induced responses, particularly when the chemically induced response above background is low. The committee recommends that the statistical approaches applied by Thomas et al. (2007) to the NTP study be applied also to the JISA study.

It is unclear whether MCL is a relevant predictor of human leukemias or other adverse health effects. Thomas et al. (2007) argue that MCL is a large granular lymphocytic leukemia (LGLL) of natural-killer (NK) cell origin that shares "some characteristics" with a rare human NK-LGLL. However, they also note that in contrast with F344 rats, human NK-LGL leukemia is rare, occurs primarily in the young, and may be associated with Epstein Barr virus (EBV) although no such virus-leukemia association is known to contribute to the etiology of rat LGLL/MCL. EPA contends that MCL is "similar" to human lymphoid cancers (T-cell and NK-LGL leukemias) and relies on a study (Stromberg 1985) that compared morphologic characteristics between rat MCL and human T-cell lymphoma. EPA considers that to be supportive evidence, despite the fact that these cancers arise in different tissues and that the cell origin in both cases is unknown. EPA states (EPA 2008, p. 4-161) that "discounting a rodent neoplasm simply because it has no human counterpart is not a scientifically defensible position. Strict site concordance is not a requirement for relevance in extrapolation of hazard potential." The committee agrees with those statements, but notes that the available data should be used to provide a more convincing argument. Similarly, EPA argues that humans are heterogeneous and so could have the same inherited susceptibility as F344 rats, but provides no scientific basis for that argument.

HUMAN STUDIES

Few human data are available for assessing the relationship between tetrachloroethylene exposure and the risk of specific cell types of lymphohematopoietic cancers. Several studies have assessed the risk of chronic lymphocytic leukemia in humans (Morton and Marjanovic 1984; Travier et al. 2002; Ji and Hemminki 2005, 2006), but otherwise the finest classification of outcomes used was "leukemia," "lymphoma," "non-Hodgkin lymphoma" (NHL), and "Hodgkin disease" (HD). The EPA draft IRIS assessment concludes (p. 4-184) that the epidemiologic data "suggested an association between lymphoma and tetrachloroethylene." The committee concurs with that conclusion but would add that the data are relatively weak and inconsistent. Associations between those cancers and exposure to tetrachloroethylene are based on very small numbers and thus are statistically unstable. The positive associations with tetrachloroethylene are sometimes observed only for lymphomas in women: NHL reported by Spirtas et al. (1991) and Anttila et al. (1995) and HD reported by Blair et al. (2003) and Miligi et al. (2006). It is not clear why those differences in sex-specific results appear; they may be due to residual confounding, in that it is unlikely that men would have appreciably lower exposures than women in the same jobs. It is also possible that sex-specific susceptibility issues are contributing to this observation. Other large cohort studies (Boice et al. 1999; Lynge et al. 2006) found no association in either women or men, and no dose-response effects have been observed. Epidemiologic studies of the association vary with study design, validity, specificity of exposure assessment, type of population studied, and sample size, all of which contribute to the inconsistency of results and reduce the committee's confidence in the conclusions that can be drawn from the data. The committee also noted a number of factual errors in this section of the IRIS draft that should be corrected; such errors detract from overall confidence in the draft's conclusions.

MODE OF ACTION

Given the high background rate of MCL in F344 rats, it is important to question whether tetrachloroethylene induces MCL or promotes an increase over the background rate. However, few data are available for addressing the question. According to EPA, a link to a mode of action (MOA) for tetrachloroethylene-induced MCL implicates a circulating genotoxic metabolite that is formed in the kidney by cleavage of a cysteine conjugate, *S*-(1,2,2-trichlorovinyl)-L-cysteine (TCVC) and may cause DNA damage in bone marrow. The EPA draft discusses studies that showed that a related (trichloroethylene-derived) cysteine conjugate, *S*-(1,2-dichlorovinyl)-L-cysteine, caused DNA alterations and toxicity in the bone marrow, lymph nodes, and thymus of calves (Bhattacharya and Schultze 1971, 1972; Lock et al. 1996). The finding that TCVC did not induce those responses in the same study does not appear to have factored into EPA's support of the hypothesis of a genotoxic MOA. The committee judges that a genotoxic MOA of tetrachloroethylene induction of MCL involving the cysteine conjugate β-lyase pathway is highly speculative and not supported by data.

The committee found some additional data on tetrachloroethylene that might be relevant for MOA analyses. They include studies by Marth et al. (Marth et al. 1985, 1989; Marth, 1987) and a study by Seidel et al. (1992) on tetrachloroethylene toxicity in mice. In the Marth et al. studies, NMRI mice were orally exposed to tetrachloroethylene at 0.05 or 0.1 mg/kg per day for 7 weeks. The mice exhibited a reversible hemolytic anemia and had microscopic evidence of splenic involvement (Marth et al. 1985), and tetrachloroethylene was found to accumulate in the spleen (as shown in Figure 2 of Marth et al. 1989), where MCL is thought to originate. Nevertheless, hemolytic anemia arises as a result of a defect in the mature red-cell membrane, as opposed to the various forms of leukemia which are thought to arise as a result of mutational changes early in bone-marrow-cell differentiation. Thus, hemolysis would not be expected to play a role in leukemogenesis. The observations reported by Marth et al. have not been reproduced or reported by any other laboratory.

Seidel et al. (1992) exposed hybrid mice (C57/BL/6 × DBA/2) to tetrachloroethylene at 270 ppm (11.5 weeks) and 135 ppm (7.5 weeks) 6 hours/day 5 days/week. Reductions in the numbers of lymphocytes/monocytes and neutrophils were observed, but they returned to control values over the next 3 weeks. There were no effects on spleen colony-forming units (CFU-Ss), but evidence of a reduction in red cells was supported by decreases in erythroid colony-forming units and erythroid burst-forming units and evidence of reticulocytosis. The data suggest a reversible bone marrow depression.

Inhibited production of both red cells and various forms of white cells have been reported after exposure to a variety of leukemogens (such as anticancer alkylating agents or benzene). The leukemogens usually decrease CFU-Ss, an effect not observed with tetrachloroethylene exposure (Seidel et al. 1992). They also usually decrease the bone marrow myeloid progenitors, CFU-GEMM, CFU-GM, and CFU-E/BFU-E, the latter of which was also decreased by tetrachloroethylene (Seidel et al. 1992). EPA should consider reviewing the evidence from models of leukemia induced in humans by chemicals (such as benzene and chemotherapeutic agents) to determine whether there are similarities with tetrachloroethylene-induced MCL.

The Marth et al. studies and the Seidel et al. study provide indirect evidence that tetrachloroethylene exposure induces effects associated with MCL and known leukemogens, respectively, but are insufficient to support the argument that tetrachloroethylene induces MCL or a related form of leukemia. In addition, those studies investigated tetrachloroethylene exposure in mice, a species in which MCL has never been observed. The only evidence that tetrachloroethylene induced MCL comes from exposure studies with F344 rats. Nevertheless, the effects of tetrachloroethylene on hemolysis in mice and on bone marrow function provide the basis of a hypothesis that could be explored to demonstrate the mechanism by which tetrachloroethylene could, within some dose range, affect the spleen.

SUMMARY

The majority of the committee finds that EPA has not adequately justified the use of MCL data over the evidence for liver or kidney cancer in its cancer risk assessment. Evidence of tetrachloroethylene-induced leukemia from epidemiologic studies is limited and inconsistent. The NTP (1986) and JISA (1993) study results of increased MCL incidences in F344 rats given tetrachloroethylene by inhalation are also questionable because of the high background rates of MCL in control animals. More thorough statistical evaluation of the data, such as the life-table analysis proposed by Thomas et al. (2007), could provide a stronger basis for drawing conclusions. However, MCL resulting from tetrachloroethylene exposure has not been observed in other strains of rats or other animal species, and no definitive evidence is available to support a hypothesized MOA by which tetrachloroethylene increases MCL in F344 rats. Those are all sources of uncertainty surrounding the relevance of MCL to human cancer risk. The information is considered in the context of the other evidence on carcinogenicity in Chapter 11, where EPA's assessment of carcinogenic risks of tetrachloroethylene is evaluated.

9

General Review of Epidemiologic Evidence Pertaining to Cancer

The Environmental Protection Agency (EPA) draft Integrated Risk Information System (IRIS) assessment of tetrachloroethylene characterizes the epidemiologic literature as supportive of classifying tetrachloroethylene as a likely carcinogen. That classification is based primarily on reported associations with hematopoietic, lymphopoietic, and esophageal cancers. There is a substantial epidemiologic literature on the potential association of exposure to tetrachloroethylene with selected malignancies. However, the committee believes that a balanced and critical review of the human epidemiologic literature provides only limited evidence that tetrachloroethylene is carcinogenic in humans. The challenges of obtaining valid estimates of exposure, in addition to the challenges inherent in observational epidemiology, make it difficult to draw conclusions about causal associations between tetrachloroethylene and cancer in humans.

The epidemiologic literature relating tetrachloroethylene to cancer is notable in three ways: a number of studies show associations with a variety of cancers, there is limited consistency between studies with respect to the associations, and few studies were able to quantify or even identify specific tetrachloroethylene exposure. The latter point, not uncommon in studies of occupational and environmental causes of cancer, makes interpretation of the literature particularly difficult. Several positive associations are reported in the literature, but the inconsistency among studies raises concern, so a consistent critical review of the literature is needed. The draft IRIS assessment does not provide the detail and methodology used for evaluating literature. Overall, it appears that the procedure was to accept the results of positive studies with little critical evaluation of validity and to dismiss null studies of similar or better methodologic rigor as flawed. If it is EPA's intention to err on the side of protecting public health when reviewing the literature, that should be stated clearly in the document. Otherwise, a clearer discussion of criteria used to identify studies of merit and a more balanced critique would strengthen the draft IRIS assessment.

The draft's critiques of studies are often uneven; studies that found no association are criticized more often than studies that found a positive association even if they had similar methodologic limitations. An example is the discussion of case-control studies on page 4-150, lines 19-31. Several of the criticized features of the case-control design that are mentioned are not inherent in the design, such as that associations may be nonlinear (this design does not require categorical exposure measures) or that duration and cumulative exposure do not address age at first exposure (this information can simply be asked of participants). Many of the studies suffered from a lack of statistical power—a common problem in studying rare cancers and exposures. However, the concern over power is uneven. On page 4-149, the absence of an association between employment in dry-cleaning and death due to lymphatic and hematopoietic cancer (Ruder et al. 2001) is attributed to lack of power. In contrast, a positive association between exposure to tetrachloroethylene and multiple myeloma in aircraft maintenance workers was based on only two deaths and is described only as noteworthy but imprecise (page 4-148, lines 6-9). There is little discussion of the potentially important limitations of proportionate-mortality studies, such as inaccuracies in death certification and the inability to adjust for potential confounders. There is some discussion of confounders in relation to the standardized mortality ratio (SMR) studies of esophageal cancer on page 4-153, but it is also unbalanced in that it focuses on adjustment for smoking but does not mention the absence of adjustment for alcohol; in addition, the effect of adjustment for smoking is derived from estimates for lung cancer and may not translate directly to esophageal cancer.

A number of errors suggest an incomplete understanding of epidemiologic and statistical methods. Such errors reduce confidence in the draft's conclusions. For example, EPA summed observed and expected cases from studies with diverse types of end points (incidence and mortality) and, using different approaches to calculating the expected values, calculated a ratio of the summed observed and expected values. Expected numbers from different studies can be added only if they are derived from the same external rates, but mortality and incidence are different. One of the most troubling misunderstandings is related to the dismissal of the results of the 2006 study by Lynge et al. In reference to that study's findings on non-Hodgkin lymphoma (and later on bladder cancer), EPA notes that exposure information was not available on about 20% of cases and of controls and that much of the exposure information came from next of kin. It then uses that to explain why Lynge et al. found no risk associated with tetrachloroethylene exposure and suggests an automatic bias toward the null due to misclassification. In the first instance, missing exposure data are analogous to nonresponse in that the subjects are not included in any classification group. Nonresponse will not introduce bias if it is nondifferential; if it is differential, it could bias an effect measure either toward or away from the null. In the second instance, exposure information from next of kin make it more likely that hazardous exposures will be overreported by the families of workers who developed cancer than by families of workers who did not; this would have resulted in overestimation, not attenuation, of the association. Similar arguments regarding the study are incorrectly made for other cancer sites, and the draft refers to the study as "uninformative." It is unclear why Lynge et al. (2006) received such critical review and papers that were methodologically less sound were accepted with little comment.

The draft IRIS assessment indicated that the strongest evidence linking tetrachloroethylene to cancer consisted of observed associations with esophageal cancer and lymphoma (page 4-184, lines 6-17). Evidence on other cancer end points—including renal, bladder, cervical, and lung cancers—is less certain and does not weigh as heavily in the assessment (page 4-184, lines 25-33). After a brief and uncritical discussion of the epidemiologic literature that references the criteria for causation outlined by Hill (1965), the document concludes that "together, the evidence on tetrachloroethylene partially fulfills several of these criteria and is suggestive of a cause and effect relationship between tetrachloroethylene and human cancer. The body of human evidence *is not sufficient* to regard tetrachloroethylene as a known human carcinogen" (p. 44-187; emphasis added). In contrast, in Chapter 6 of the draft ("Characterization of Hazard and Dose-Response"), the evidence associating tetrachloroethylene exposure with cancer is stated more confidently (page 6-5, lines 31-35; page 6-6, lines 1-5; page 6-10, lines 27-29 and 31-35; and page 6-11, lines 1-6). It is difficult to reconcile the discussion in Chapter 4 with the conclusion in Chapter 6.

ESOPHAGEAL CANCER

The draft IRIS assessment emphasizes the association between tetrachloroethylene and esophageal cancer primarily because of the results of three studies: by Vaughan et al. (1997), Ruder et al. (2001), and Blair et al. (2003). The work by Blair et al. and Ruder et al. were mortality studies of dry-cleaner union members, and the latter was a community-based case-control study. It is interesting to compare the results of the two studies. With the same methods, the populations were enumerated from similar sources and followed for similar periods. Blair et al. followed 5,369 union members in St. Louis who worked for at least 1 year during 1948-1993. The population studied by Ruder et al. included 1,708 workers selected from union rosters in California, Illinois, Michigan, and New York. Both studies reported an excess risk of death from esophageal cancer; Blair et al. reported an SMR of 2.2 (95% confidence interval [CI], 1.5-3.3) and Ruder et al. an SMR of 2.47 (95% CI, 1.35-4.14). The excess in the paper by Ruder et al. was limited to workers with at least 20 years since first employment and was highest in those with at least 5 years of exposure (SMR, 5.03; 95% CI, 2.41-9.47). Blair et al. reported similar SMRs in workers with little or no exposure (SMR, 2.1; 95% CI, 0.9-4.4) and those with medium or high exposure (SMR, 2.2; 95% CI, 1.1-3.5).

Esophageal cancer is also associated with smoking and alcohol consumption, which are difficult to control for in mortality studies because the data are often not available. The studies of Blair et al. and Ruder et al. also reported an excess of deaths from other causes associated with smoking, including lung cancer, emphysema, and heart disease. EPA's draft IRIS assessment discounts potential confounding by smoking but does not adequately support its conclusion in the section on esophageal cancer (page 4-153, lines 30-33). In contrast with the findings of Blair et al. and Ruder et al., a large mortality study (Boice et al. 1999) in a population of aircraft manufacturers (N = 77,965) had an appreciable number of workers with routine (N = 2,631) and intermittent (N = 3,199) exposure to tetrachloroethylene but reported no association between that exposure and esophageal cancer. The case-control study by Vaughan et al. (1997) reported an increased but not significant odds ratio (OR) for dry-cleaning work, which was adjusted for smoking habit and alcohol consumption. That estimate was based on only two exposed cases, however, and, particularly when multiple covariates were adjusted for, was too statistically unstable to be informative (OR 3.6; 95% CI, 0.5-27.0). A methodologically sound nested case-control study by Lynge et al. (2006) reported no association between working as a dry-cleaner and esophageal cancer. Those negative findings were dismissed by EPA because some of the population could not be classified by exposure. As discussed earlier, this does not preclude the use of results based on subjects on whom exposure data were available.

Overall, there is limited evidence to support an association between tetrachloroethylene and esophageal cancer. The two mortality studies of dry-cleaners are suggestive of an association, but the potential for confounding by smoking and alcohol consumption is appreciable. Thus, the committee therefore concluded that the epidemiologic literature is not sufficient to support an association between tetrachloroethylene and esophageal cancer.

LYMPHOID CANCERS

EPA's draft IRIS assessment concludes that the epidemiologic data "suggested an association between lymphoma and tetrachloroethylene" (p. 4-184). The committee concurs with that conclusion but adds that the data are relatively weak and inconsistent. The rationale for the committee's conclusion is discussed in detail in Chapter 8.

Epidemiologic studies of the association between exposure to tetrachloroethylene and lymphoid cancers vary in design, validity, specificity of exposure assessment, type of population studied, outcome, and sample size, all of which contribute to the inconsistency of results and reduce confidence in conclusions that are drawn from the data.

OTHER CANCERS

A number of studies have reported associations between tetrachloroethylene and other cancers, including cervical, lung, and bladder cancer. The results for those cancers are less consistent but should not be dismissed. The draft IRIS assessment considered those end points but did not weigh them heavily in the classification of tetrachloroethylene as a human carcinogen. That is appropriate.

GENERAL COMMENTS ON THE ENVIRONMENTAL PROTECTION AGENCY'S PRESENTATION OF EPIDEMIOLOGIC EVIDENCE ON CANCER

One of the biggest difficulties in assessing the cogency of the EPA's assessment related to cancer is how the data are organized in the tables and some parts of the text. It would be much easier to evaluate the overall picture of results regarding tetrachloroethylene and a particular cancer if the tables were organized by cancer type as opposed to the current format, which organizes them by study design. The current format requires the reader to jump between sections for cohort mortality, incidence, and case-control studies. Studies are sometimes further categorized as to the type of worker included (for example, dry-cleaner vs degreaser); this makes it extremely difficulty to evaluate the overall consistency or lack of consistency in results related to specific cancers.

Errors in reporting results also occur occasionally. For example, the draft reports (on page 4-150, lines 1-3), in relation to Hodgkin disease, "a statistically significantly elevated risk for male [sic] with a job title of dry cleaner or laundry worker (Costantini et al. 2001)." The result from Costantini et al. for that group in relation to Hodgkin disease was an OR of 2.5 (95% CI, 0.3-24.6), which is not significant and was based on a single case.

The overall impression is that data are presented to support a positive association between tetrachloroethylene and cancer and that studies that found no such association are criticized or minimized. EPA should provide a clearer discussion of criteria used to identify studies of merit and a more balanced critique to strengthen the draft IRIS assessment.

RESEARCH RECOMMENDATIONS

Population-based studies, preferably in well-defined occupational cohorts, that can measure both cancer incidence and mortality and have sophisticated exposure reconstruction components that are specific to tetrachloroethylene would add significantly to the literature. The studies must also be adequately controlled for the effects of smoking and alcohol consumption to address the lingering questions of the association between tetrachloroethylene and esophageal cancer. In the absence of data to control for these confounders, sensitivity analyses should be conducted to estimate the exposure effect after adjustment under reasonable sets of assumptions regarding smoking prevalence and the strength of smoking effects. Further research that classifies exposure only by occupational title will not add to the literature.

10

Reference Values for Tetrachloroethylene

The Environmental Protection Agency (EPA) draft Integrated Risk Information System (IRIS) assessment of tetrachloroethylene provides the agency's assessment of the potential human health effects of exposure to the chemical. For noncancer effects, EPA proposes to establish an oral reference dose (RfD) and an inhalation reference concentration (RfC), which the agency defines as estimates (with uncertainty spanning perhaps an order of magnitude) of a daily oral exposure and a continuous inhalation exposure, respectively, of the human population (including sensitive subgroups) that are likely not to pose an appreciable risk of deleterious effects during a lifetime. The proposed RfC for tetrachloroethylene is 0.016 mg/m^3, and the proposed RfD is 0.004 mg/kg per day. This chapter discusses how those reference values were determined.

SELECTION OF CRITICAL END POINT AND STUDIES

EPA selected neurotoxicity—specifically, outcomes of visual and cognitive dysfunction—as the critical noncancer health effect of tetrachloroethylene. As described in Chapter 3, epidemiologic and human studies involving controlled exposures have provided evidence of those effects. The experimental-animal literature available when the draft IRIS document was written also provided strong evidence that tetrachloroethylene is neurotoxic. One study (Mattsson et al. 1998) offered support of effects on visual function. New studies have provided further support of effects on the visual system and signal-detection tasks (Oshiro et al. 2008; Boyes et al. 2009) in animals. Although the committee supports EPA's decision to use neurotoxicity as a critical end point, it recommends more focused assessments of specific criteria related to study design and methods as part of the process of selecting critical studies for deriving reference values.

The committee found that EPA reviewed all the relevant studies available at the time that the draft was written and agrees with many of the limitations that are noted, beginning on page 4-101. The committee also found, however, that the draft sometimes failed to consider weaknesses in study methods or inconsistencies in results, two factors that should carry great weight in selecting key studies for calculating an RfC. For example, test outcomes (neurologic signs, emotional lability, choice reaction time, cancellation d2, and digit symbol) in a study by Seeber (1989) were worse in the low-exposure group compared with the high-exposure group. EPA's discussion of the study (Section 4.6.1.2.2) did not mention that discrepancy. In another example, the committee judged the study by Echeverria et al. (1995) to be stronger than is characterized in the draft assessment (see detailed discussion in Chapter 3 of the present report). EPA discounted the study because (p. 4-77 to 4-78) "the lack of an unexposed control group limits the ability of the study to fully characterize the magnitude of the effects on visuospatial ability and to detect exposure-related symptoms or effects on tests of non-visuospatial cognitive ability. It also limits the extrapolation of the results to other populations exposed to tetrachloroethylene." The committee judged that although there was no unexposed comparison group, the use of an internal comparison group (the group with the lowest exposure) has the advantage that any selection and confounding factors related to working in dry-cleaning facilities are present in both groups and reduces potential confounding by unmeasured factors.

The committee applied several criteria in selecting the epidemiologic studies that it considered most useful in establishing reference values for tetrachloroethylene. Three general criteria were addressed: the validity of individual studies, the internal consistency of results with the hypothesis of a causal role for tetrachloroethylene (for example, is there an association in a low-exposure group but not in the high-exposure group?), and the consistency of the findings with what is known from other sources (how the study fits into the overall picture of what is known). Those criteria are discussed in detail in Chapter 3.

EPA selected the study by Altmann et al. (1995), conducted in Mülheim, Germany, for calculating the RfC because it involved environmental exposures that are more relevant than occupational exposures for determining values designed to protect public health and it used a standardized computer-assisted testing battery. Those study factors are reasonable considerations, but they are not the most relevant for selecting a critical study. The committee concluded that the validity of the 1995 Altmann et al. study was seriously compromised by methodologic deficiencies, which are discussed in detail in Chapter 3 and summarized briefly below.

- The most important concern is that the referent group was inappropriate in that it did not represent the counterfactual example. It was selected from among employees of the Public Health Office or the Medical Institution of Environmental Hygiene in Mülheim and matched to exposed subjects by age and sex. This selection bias resulted in a reference group clearly was more educated than the exposed group, and because the authors used only three categories of education, it is unlikely that differences in education were adequately controlled for. Because several of the primary outcomes are influenced by education, it is likely that substantial confounding remained. For example, there was no association between tetrachloroethylene and visual evoked potentials (VEPs). That is important because visual deficits have been the most consistently reported effects of tetrachloroethylene, and they are outcomes that are essentially unrelated to education. Measures of vigilance, attention, and visual memory are strongly associated with education, and the exposed group had poorer performance in them, whereas measures of eye-hand coordination and finger tapping, which are weakly related to education, were similar in the two groups.

- The rationale for the selection of study participants was poorly described, and several of the exposure measurements in those supposedly exposed were not reported. Without that information, it is impossible to determine whether this was a biased sample (that is, whether others were excluded for reasons other than study design).
- Tetrachloroethylene was measured in air samples from homes for 7 days. Figure 1B of the paper is supposed to show indoor air concentrations for exposed participants and referents, but no concentrations are shown for the referent group. The amount of time that residents spent in their apartments is unknown. Time out of the apartments before neurobehavioral testing was unknown but was believed to account for lower blood concentrations of tetrachloroethylene before testing.
- In the analyses, exposure was defined by group membership (yes-no variable) rather than by markers of exposure. Therefore, no dose-effect relationship was established in the exposed group. As stated above, group differences in neurobehavioral performance were more likely to be related to residual confounding by education or pre-exposure intellectual capacity.
- The Neurobehavioral Evaluation System battery used to assess brain dysfunction related to exposure appropriately included four subtests that have been shown to be associated with solvent exposure in other research. However, the battery has no norms in this population, and some of the tests have not been well validated with regard to what they reveal about brain damage from any cause. The absence of norms makes it especially important to have basic, standardized measures of intellectual function that can be used to characterize the longstanding cognitive abilities (native intellectual capacity) of the two groups so that differences between the groups can be correctly attributed to exposure.

On the basis of the study selection criteria noted earlier—which emphasized validity, methodology, and consistency with the literature—the human studies that the committee judged most appropriate to use as points of departure for derivation of the RfC are Altmann et al. (1990), Cavalleri et al. (1994), Gobba et al. (1998), and Echeverria et al. (1995). The details of those studies and the reasons for their selection are discussed fully in Chapter 3 and summarized briefly here. The study by Altmann et al. (1990), who used controlled exposures in an experimental chamber, was chosen because it used random assignment to exposure groups, which reduced the potential for confounding of any associations between exposure and outcomes, and the exposure dosage was known. The study by Cavalleri et al.

(1994) was useful because it examined an occupational cohort of 33 dry-cleaners and it included followup assessments 2 years later, as reported by Gobba et al. (1998). Some members of the cohort continued to be exposed to the same workplace concentrations of tetrachloroethylene, and others worked in facilities where exposures had been reduced. The 1998 study by Gobba et al. was useful in that it allowed assessment of color vision before and after alterations in workplace exposure to tetrachloroethylene and because exposure concentrations could be estimated. The primary advantages of the study by Echeverria et al. (1995) were the reduction in potential confounding and confounding due to the use of an internal referent group and the ability to examine exposed workers for a dose-response effect with respect to measures of visuospatial performance on the basis of estimated cumulative lifetime exposure to tetrachloroethylene.

Among the animal studies considered by the committee, the one by Boyes et al. (2009) was judged to be appropriate to use as a point of departure for derivation of the RfC. The most sensitive end point in the study was the F2 (frequency-doubling) component of the evoked potential spectrum, a measure thought to reflect the activity of cortical neurons that respond to both stimulus offset and onset. The investigators also conducted a toxicokinetic analysis relating exposure concentration and duration to brain concentration. From that analysis, brain concentrations of tetrachloroethylene were linked to visual function.

DOSE METRICS

With respect to neurotoxicity, EPA's use of the blood area under the curve (AUC) for tetrachloroethylene with various routes of exposure appears to be justified. The physiologically based pharmacokinetic (PBPK) model simulations presented in Figures 3-4, 3-5, 3-6a, and 3-6b of the draft IRIS assessment (EPA 2008) do suggest that the three PBPK models collectively describe the variation in blood and exhaled-breath concentrations of tetrachloroethylene observed in controlled human exposures. That provides confidence that later calculations of the tetrachloroethylene AUC during various exposure scenarios are accurately captured. A better dose metric for use in the neurotoxicity assessment might be the AUC for tetrachloroethylene in the brain. However, given the rapid partitioning of tetrachloroethylene between blood and well-perfused tissues and the lack of experimental data on brain tetrachloroethylene concentrations, the use of the blood AUC as a surrogate was appropriate. (As noted in Chapter 3, there are now some data that might be used in developing PBPK models of brain concentration.)

ROUTE-TO-ROUTE EXTRAPOLATION

EPA has chosen to use the venous-blood AUC as the route-to-route dose metric for extrapolating an inhalation exposure of tetrachloroethylene to a corresponding oral equivalent dose. The rationale behind that approach is sound and adequately explained in the draft. However, the implementation of the approach raises serious methodologic concerns related to inappropriate use of the selected PBPK models and uncertainties in the fraction of an oral tetrachloroethylene dose that is metabolized. All three of the selected PBPK models were formulated and validated specifically against inhalation exposures. There was no attempt to validate model predictions against blood tetrachloroethylene concentrations after oral dosing. To use the PBPK models, EPA has empirically assumed a value for the rate of oral absorption of tetrachloroethylene, which is entered as a constant in the models. That approach is inferior to direct estimation as was used in other published PBPK models, such as the Gearhart et al. (1993) or Dallas et al. (1995) models (the latter only for rats and dogs). The latter PBPK models might have been better choices to begin this exercise. Better still, a harmonized PBPK modeling approach to synthesize important aspects of the various models into a single model would have provided the greatest confidence in the route-to-route extrapolation. See Chapter 11 for further discussion of the limitations of the PBPK modeling and the proposal to develop a harmonized model.

CHARACTERIZATION OF UNCERTAINTIES

The committee reviewed EPA's application of uncertainty factors in deriving sample reference values on the basis of different studies. It found that the narrative made it clear what uncertainty factors were used but that there were some instances in which a supporting rationale was not provided for departure from the default option and other instances in which departures from the default option should have been considered.

Extrapolation from Lowest Observed-Adverse-Effect Level to No-Observed-Adverse-Effect Level

A factor of 10 was used consistently by EPA when a lowest observed-adverse-effect level (LOAEL) from a study was used instead of a no-observed-adverse-effect level (NOAEL). That is consistent with EPA policy. A benchmark dose (BMD) can be treated as a NOAEL, but no studies of neurotoxicity that could support a BMD calculation had been published when the draft was written. More recent studies of neurotoxicity would support such a calculation (Oshiro et al. 2008; Benignus et al. 2009; Boyes et al. 2009).

Extrapolation from Animals to Humans

The uncertainty factor for extrapolating animal data to humans is considered to have toxicokinetic and toxicodynamic aspects. EPA judged that an uncertainty factor of 3 was adequate to address these uncertainties. EPA applied that approach consistently, but the rationale for doing so was not adequately described. Specifically, the draft cites an EPA (1994) document, but it would have enhanced transparency if it summarized briefly why an uncertainty factor of 3, rather than the default factor of 10, was used.

Human Variation

The application of a default factor of 10 to account for interindividual variation is justified because of the paucity of data on sensitive populations, including developing and aging organisms. Its use is appropriate and in accordance with EPA guidance.

Extrapolation from Subchronic Exposure to Chronic Exposure

The criteria for selecting the value of the uncertainty factor for extrapolating from subchronic exposure to chronic exposure were not clear, and this uncertainty was handled inconsistently in the draft IRIS assessment. It was noted (p. 5-13) that "a factor to address the potential for more severe toxicity from chronic or lifetime exposure to tetrachloroethylene is not used in this assessment. The epidemiologic studies, except for Schreiber et al. (2002), are all of median duration of exposures of more than 15% of a 70-year lifespan. There are no data to suggest that continuing exposure to tetrachloroethylene can increase the severity of effects; duration-response trends are not generally evident in the human studies." On the basis of that rationale, no uncertainty factor for extrapolating to lifetime exposure was applied to the Altmann et al. (1995) study. However, in the discussion of the studies that support the RfC, a factor of 10 was applied to the Schreiber et al. study of day-care workers even though the mean exposure period was said to be 4 years, during which 23% of the time would involve exposure. It is not clear how that pattern would differ from residential exposure of people who work outside the home during the day. More directly, however, if EPA believes that longer exposures do not increase neurotoxicity or, by implication, that shorter exposures do not diminish it, one may question why a factor of 10 is applied to the results of the Schreiber et al. (2002) occupational study but not to the results of other occupational studies. Overall, the committee found that the literature provides little information about the possibility of cumulative toxicity from chronic exposure to tetrachloroethylene. No animal studies of chronic, life-long exposure were located, and except for Gobba et al. (1998) the epidemiologic studies did not involve long-term followup.

There is inconsistent use of the uncertainty factor when the sample reference value is based on the results of animal studies—Mattsson et al. (1998) and Rosengren et al. (1986). A factor of 10 was applied to the Mattsson et al. results and a factor of 3 to the Rosengren et al. results even though the two studies were of similar duration. EPA's rationale (p. 5-15) was that "a subchronic to chronic factor of 3, rather than 10, was applied for Rosengren et al. (1986) in light of the large overall uncertainty for this study associated with extrapolating from a

LOAEL to NOAEL, from animal to humans, for human variation, and for database deficiencies; the total uncertainty factor was 3,000." That justification is not clear. The reason for modifying the uncertainty factor may be that it is EPA's *policy* to limit the overall uncertainty to 3,000 in deriving RfCs (EPA 2002). If that is the reason, it should be stated explicitly. If not, better justification should be provided.

The committee believes that an uncertainty factor of 3 should have been considered for animal studies like that of Mattsson et al. (1998) in which exposure occurred for 6 hours/day 4 days/week for 13 weeks. If that exposure regimen is treated in the same manner as acute exposure by applying a higher factor, doing so should be justified. Some discussion of the issue would improve the draft IRIS assessment.

Database Deficiencies

In the derivation of RfCs on the basis of neurotoxicity, EPA used a factor of 3 for database deficiencies because of the inadequacy of the experimental literature designed to characterize hazard and dose-response. Key deficiencies identified were inadequate data to address childhood or other life-stage susceptibility, a paucity of animal studies (especially studies of developing animals and of chronic, low-level exposures) designed to investigate neurotoxicity or to define and characterize dose-response relationships, and inadequate database on cognitive testing. It was unclear whether a factor of 3 was adequate to address these uncertainties because there was some overlap with the factor of 10 applied for human variation, which also addressed developmental concerns.

The committee recommends that EPA revisit and defend more clearly its decision to apply a factor of 3 for database deficiencies in light of new data and the committee's findings in Chapter 3. New studies include, for example, recent papers from researchers in EPA's National Health and Environmental Effects Research Laboratory provide excellent data from well-designed studies using controlled, acute exposures that link deficits in visual function and signal detection with atmospheric tetrachlorethylene concentrations and instantaneous concentrations in the brain. This includes papers by Oshiro et al. (2008) and Boyes et al. (2009) investigating function and by Shafer et al. (2005) on mechanisms, which is described in the IRIS document but not fully integrated. These studies link neural or behavioral effects to actual brain concentrations of tetrachloroethylene or to their estimated concentration using PBPK modeling. Thus, the animal literature on controlled acute exposure is now stronger. Notable gaps in the animal literature still include the paucity of studies of developmental or chronic exposures. Another consideration is that the committee found the human study of exposed children (Schreiber et al. 2002) to be methodologically flawed. The committee judges these to be serious gaps in the database, which suggests that a factor of 3 may be inadequate to account for database deficiencies.

Human Equivalences

Human equivalences are said to reflect adjustments from a less than continuous exposure to continuous exposure, such as might occur in a residence. That assumes continuous exposure although few people are in their homes 24 hours/day. The human equivalence factor is supposed to involve an adjustment from exposures 5 days/week to exposures 7 days/week by multiplying by 5/7 or from 8 hours/day to 24 hours/day when experimental exposures (as in animal studies) are less than continuous. For human occupational exposures, a 10/20 factor is applied to accommodate an increased respiration rate during work; however, when this factor is applied, the adjustment to a 24-hour day is not applied, but the adjustment to a 7-day week is. For oral exposures but apparently not for inhalation exposures, there is an allometric adjustment for body-weight differences. Those considerations are in accord with EPA policy but are far from intuitive and should be summarized in the document where they are applied in the tables. The draft's Figure 5-7 clearly presents that approach in estimating cancer risk, but the figure does not apply to risk posed by inhalation. It is sometimes difficult to see how the "human equivalence" factor is determined for a particular study, and some rationale for its calculation would increase understanding of EPA's approach. For example, the adjustment for the Mattsson et al. study is not described, but it appears to be an adjustment from exposures 6 hours/day 5 days/week to 24 hours/day 7 days/week.

GRAPHICAL PRESENTATION OF REFERENCE VALUES

The draft IRIS assessment provides graphical presentations of noncancer reference values for tetrachloroethylene (Figures 5-1, 5-2, and 5-4). One figure (Figure 5-1) illustrates reference values based on different neurotoxicity studies and two figures (Figures 5-2 and 5-5) compares EPA's selected reference value based on neurotoxicity with other reference values based on other noncancer effects. The committee strongly supports the use of such graphical aids. In general, the approach is intended to make it clear which uncertainty factors were applied, to which studies they were applied, and the effects of particular assumptions. However, the figures in the draft document fail to accomplish that. The shading used in the legend for the figures does not match the shading in the figures, so it is impossible to determine which uncertainty factors were used. By including a small number of studies, the figure on neurotoxicity sample reference values (Figure 5-1) also misses an opportunity to show the degree to which the literature converges on a limited range of sample values. Convergence of estimated values from studies that are methodologically sound, even if they are not listed as key, would support the RfC proposed by EPA.

To synthesize the literature, the committee considered a graphical approach that shows how sample reference values that might be derived from the

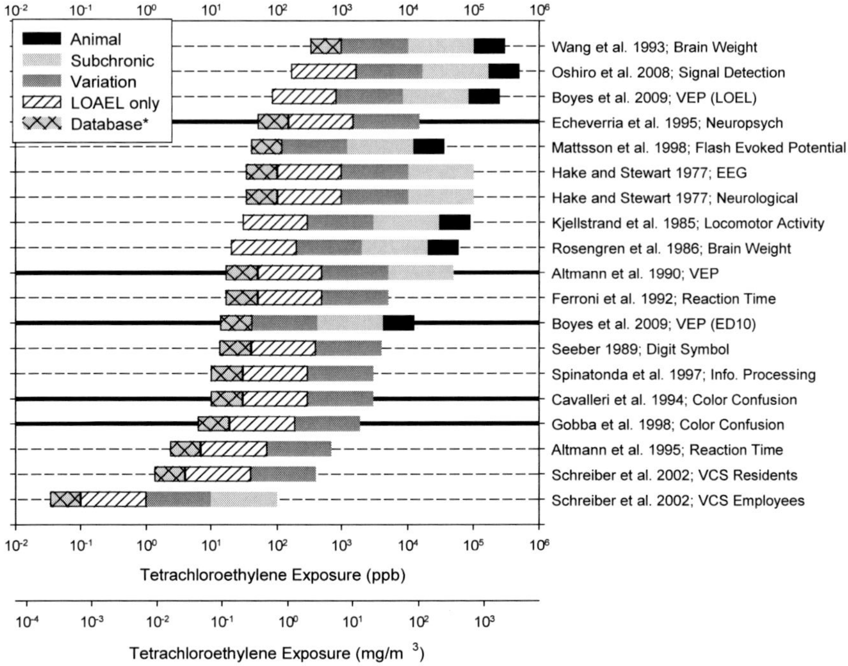

FIGURE 10-1 Distribution of sample reference values. Each horizontal bar represents a single study. Thick, horizontal lines represent studies identified by the committee as most applicable to the development of an RfC. The right end of a bar is at the "point of departure" and is based on concentrations used in the referenced study after conversion to "human equivalencies" or, in the case of animal studies, after adjustment for continuous exposure. Uncertainty factors are illustrated in different shadings: a factor of 3 if it is necessary to extrapolate from animals to humans (black); a factor of 10 if it is necessary to extrapolate from acute or subchronic exposure to chronic exposure (light gray); a factor of 10 for individual variation to account for sensitive individuals (dark gray); a factor of 10 if the study did not contain a NOAEL (diagonal lines) and a factor of 3 for uncertainty in the data base as applied by EPA (light gray, cross-hatched). *A maximum total uncertainty factor of 3,000 was applied for the purpose of this exercise. Where this might be exceeded, the maximum was achieved by omitting the "database" uncertainty so that other uncertainties could be visualized. The committee has recommended that EPA review the uncertainty factors to ensure that they are appropriately explained and used consistently, so some of the individual values used here could be subject to change. In some cases, EPA might judge that the total uncertainty exceeds 3,000 and would, therefore, not use that study to derive a sample reference value. Source: Graphic developed by M. Christopher Newland.

different studies of neurotoxicity compare with one another (see Figure 10-1). That was done by using the studies described in the draft and two studies (Oshiro et al. 2008; Boyes et al. 2009) published since the draft was written. The approach enables the visualization of the range of concentrations studied, the identification of clusters of studies, and the isolation of especially low or high reference value estimates that might be derived from a particular study. The figure includes seven data points derived from animal studies (identified by a black bar on the right), three from controlled human exposures (identified by a light gray bar and the absence of a black bar), and studies of environmental or occupational exposures. The convergence of sample reference values into clusters would confer confidence on the use of a critical study if other studies led to similar conclusions.

The points of departure for the pre-2004 studies came from Tables 4-4 and 5-2 of the draft document, so they were human adjusted equivalent concentrations or, in the case of animal studies, adjustments for continuous exposures as appropriate. Uncertainty factors were based on how they are typically applied (see pp. 5-11 onward in the draft) even when the committee disagreed with their application. For example, the committee recommends an uncertainty factor different from that applied by EPA for the Schreiber et al. (2002) study. EPA applied an uncertainty factor of 10 to the Schreiber et al. results to extrapolate from "subchronic to chronic exposure," but the study involved long-term environmental exposure of day-care workers, so the committee believed that this uncertainty factor not necessary. EPA's factor of 10 was retained in the graphical display, and the RfD calculated for occupational exposure by using this factor appears unusually low.

Studies published since the EPA draft was written are also included in Figure 10-1. One study (Oshiro et al. 2008) identified a LOAEL of 500 ppm (acute). The dependent measure was a signal-detection task. Uncertainty factors for the study would include a factor for extrapolation from animals to humans (3), one for extrapolation from acute to chronic (10), one for sensitive populations (10), one for absence of a NOAEL (10), and the routine one for database uncertainties (3)—for a total uncertainty factor of 9,000. For the purposes of this exercise to show the full database, a maximum total uncertainty factor of 3,000 was applied. The committee has recommended that EPA review the uncertainty factors to ensure that they are appropriately explained and used consistently. In some cases, EPA might judge that the total uncertainty exceeds 3,000 and would, therefore, not use that study to derive a sample reference value. In the graph, the total uncertainty factor was reduced to 3,000 by not showing the uncertainty factor of 3 for database uncertainties. The second study (Boyes et al. 2009) reported a LOAEL of 250 ppm for VEPs evoked by a grid of vertical bars; this is similar to the contrast-sensitivity task. The same factors used in the Oshiro et al. study would be applied here, so a total uncertainty factor of 3,000 (similarly reduced from 9,000) was applied to calculate a sample reference value of 83.3 ppb. The Boyes et al. study also used PBPK modeling to estimate the shape of the relationship between brain concentration and VEP. An ED_{10}, the brain concentration that produced a 10% change in VEP (the last figure in the Boyes et al. paper), was estimated. To estimate a point of departure from the ED_{10}, the exposure concentrations, in parts per billion, that would produce this brain concentration were estimated by back-calculating from relationships between brain and atmospheric concentrations in the authors' Figures 1 and 2. An ED_{10} of 0.687 mg/mL comes from their Figure 7. From Figure 1, it can be estimated that the brain:blood ratio is 33:12 (at peak), so 0.687 mg/L in brain corresponds to 0.25 mg/L in blood. From Figure 2, it can be estimated that 50 ppm in air corresponds to a peak (and near asymptote) of 1 mg/L in blood. Therefore, the blood tetrachloroethylene concentration of 0.25 mg/L should result from 12.5 ppm in air. The committee recognized that those are rough estimates that assume linearity and that a more precise estimate could be obtained with modeling. The estimate is included here only as an illustration. The 12.5-ppm point of departure yields a sample reference value of 14 ppb after application of uncertainty factors for extrapolation from animals to humans (3), acute exposure (10), variation in sensitivity (10), and database uncertainty (3). It is unclear whether an uncertainty factor should be applied for the absence of a NOAEL in the study.

Some observations can be made from the figure. The majority of sample reference values are centrally clustered, but there is a wide spread to both the lower and higher ends. Although the overall range of the 19 sample values is 0.03-333 ppb (0.0002 - 2.6 mg/m^3), it is reduced to about 6 to about 50 ppb (0.04 - 0.34 mg/m^3) when restricted to the five strongest studies. EPA's RfC of 2 ppb (0.016 mg/m^3) calculated on the basis of the Altmann et al. (1995) study falls below the range and is higher than only the two other human studies, which were conducted by Schreiber et al. The Altmann et al. (1995) and Schreiber et al. (2002) studies are discussed and critiqued elsewhere, where it is noted that the makeup of the critical comparison groups is confounded and that this makes it difficult to attribute differences seen in dependent variables to tetrachloroethylene. The figure enhances transparency by showing how studies converge on a range of reference value estimates and how the study or studies selected as the one(s) to be used for establishing the final RfC compares with other studies.

The three studies that yield sample reference values above 50 ppb are the ones that identified effects at relatively high exposure concentrations because the end points were relatively insensitive or, like in the Oshiro et al. (2008) study, are of very high quality but used high exposure concentrations, so that a low end of the dose-effect curve was not readily identifiable. While the Boyes et al. (2009) study is considered a critical one by the committee, the sample reference value based on the LOAEL from the study (as opposed to the ED_{10}) was considered to have too much uncertainty associated with it to be used as a point of departure. The consistency in the middle ranges among epidemiologic studies and controlled-exposure human studies, as well as in animal studies, provides support for points of departure in these ranges. Despite the use of different exposure regimens and concentrations among animal studies, human chamber studies, and occupational and environmental studies, there is a reasonable coherence in the sample reference values. Finally, to keep the maximum uncertainty factor to 3,000, the "database" factor of 3 was omitted from four estimates for the purposes of the exercise in Figure 10-1. Reinstating this factor would not substantively change the conclusion about the consistency in reference concentrations.

The graphical display in Figure 10-1 shows a distribution of sample reference values based on neurotoxic effects observed in epidemiologic studies, controlled human experiments involving healthy volunteers, and animal experiments involving different species. Exposure ranged from chronic to acute. The studies involved different neurotoxic end points that are differentially sensitive to tetrachloroethylene exposure. Whereas uncertainty factors applied to a point of departure adjust uncertainties specific to their corresponding studies, the collective distribution of reference values provides an overarching measure of uncertainties, weight of evidence, sensitivities, and other sources of variation among different studies.

This approach could also be applied to EPA's other graphical presentations of reference values based on other noncancer end points. Such an approach would allow organ-specific reference values to be put in context with one another. For example, the degree to which sample reference values for an organ system cluster, or fail to do so, might be viewed as evidence of the degree to which different studies tap similar toxic mechanisms, kinetics, end points, or other important characteristics of a chemical.

11

Cancer Risk Estimates for Tetrachloroethylene

The Environmental Protection Agency (EPA) draft Integrated Risk Information System (IRIS) assessment of tetrachloroethylene provides the agency's assessment of the potential human health effects of exposure to the chemical. For cancer, EPA provides a characterization of the weight of evidence of human carcinogenicity and quantitative estimates of inhalation unit risks and oral slope factors. A unit risk is the upper bound of excess lifetime cancer risk estimated to result from continuous exposure to an agent at a concentration of 1 $\mu g/m^3$ in air. An oral slope factor is the upper bound, approximating a 95% confidence limit, of the increased cancer risk from lifetime exposure to an agent; it is usually expressed as a proportion (of a population) affected per milligram per kilogram per day. For tetrachloroethylene, EPA proposes a range of inhalation unit risks of 2×10^{-6} to 2×10^{-5} per microgram per cubic meter and a range of oral slope factors of 1×10^{-2} to 1×10^{-1} per milligram per kilogram per day. These ranges reflect the application of three physiologically-based pharmacokinetic (PBPK) models to the same data. This chapter discusses how those cancer risk estimates were determined by EPA.

CANCER CLASSIFICATION

EPA asked for an evaluation of whether conclusions it has drawn in the draft IRIS assessment are consistent with its cancer guidelines (EPA 2005a), specifically with regard to its characterization that tetrachloroethylene is "likely to be a human carcinogen by all routes of exposure." Box 11-1 presents EPA's guidelines for determining such a classification.

The committee considered those guidelines, and guidelines for the other descriptors, and concluded that EPA has documented that its conclusion has been drawn from the results of bioassays that found increased incidences of hepatocelluar tumors, hemangiosarcomas, mononuclear-cell leukemia (MCL), and renal tumors in laboratory animals and to a lesser extent from epidemiologic evidence. EPA's decision to characterize tetrachloroethylene as likely to be a human carcinogen as opposed to "carcinogenic to humans" appropriately reflects that there could be deficiencies or potential inaccuracies in interpretation of the data. Some of those possible deficiencies and inaccuracies are discussed below for each of the data sets.

BOX 11-1 EPA Cancer Guidance for Concluding a Chemical Is Likely to Be Carcinogenic to Humans (EPA 2005)

This descriptor is appropriate when the weight of evidence is adequate to demonstrate carcinogenic potential to humans but does not reach the weight of evidence for the descriptor "Carcinogenic to Humans." Adequate evidence consistent with this descriptor covers a broad spectrum. As stated previously, the use of the term "likely" as a weight of evidence descriptor does not correspond to a quantifiable probability. The examples below are meant to represent the broad range of data combinations that are covered by this descriptor; they are illustrative and provide neither a checklist nor a limitation for the data that might support use of this descriptor. Moreover, additional information, e.g., on mode of action, might change the choice of descriptor for the illustrated examples. Supporting data for this descriptor may include:

- an agent demonstrating a plausible (but not definitively causal) association between human exposure and cancer, in most cases with some supporting biological, experimental evidence, though not necessarily carcinogenicity data from animal experiments;
- an agent that has tested positive in animal experiments in more than one species, sex, strain, site, or exposure route, with or without evidence of carcinogenicity in humans;
- a positive tumor study that raises additional biological concerns beyond that of a statistically significant result, for example, a high degree of malignancy, or an early age at onset;
- a rare animal tumor response in a single experiment that is assumed to be relevant to humans; or
- a positive tumor study that is strengthened by other lines of evidence, for example, either plausible (but not definitively causal) association between human exposure and cancer or evidence that the agent or an important metabolite causes events generally known to be associated with tumor formation (such as DNA reactivity or effects on cell growth control) likely to be related to the tumor response in this case.

Mononuclear Cell Leukemia

An increased incidence of MCL in F344 rats has been reported in two bioassays. The biological significance of these increases was debated by the committee because increases were observed only in one strain of rat, which is known to have a high background incidence of MCL. Control data on F344 rats indicate background rates of MCL ranging from 7.9-52.5% in males and 2.1-24.2% in females (Thomas et al. 2007), which make it difficult to interpret the biological significance of increases observed in two studies from different laboratories (NTP 1986; JISA 1993) because of the lack of information on mode of action. Statistical methods, such as survival data analysis, which incorporate data from multiple dose groups simultaneously for dose-response analysis rather than pair-wise comparison should be explored to aid in interpretation. For example, as noted in Chapter 8, Thomas et al. (2007) have made a case that using life-table analysis to examine the MCL data provide an improved approach for interpreting the significance of a dose-response for a possible carcinogenic effect. They judged that there was a positive association between tetrachloroethylene and MCL in the NTP study when such criteria were applied, but recommended a weight-of-evidence evaluation be performed before drawing conclusions. The committee observed that the data showed inconsistency in statistical significance between sexes and uncertainty about the shape of the dose-response curve, especially in the lower range of the NTP study. There is some support from epidemiologic studies which suggest an association between lymphoma and tetrachloroethylene, but the data were relatively weak and inconsistent. A difficulty with interpreting the findings is differences in opinion about the biological concordance between MCL and human lymphohematopoietic cancer. Some members judged that similarities between human natural killer large granular lymphocyte leukemia and rat MCL, as well as mechanistic studies the committee recommended be added to EPA's assessment, are adequate to assume human relevance, whereas other believe more research is needed to establish the relevance. In addition, there was little information on a mode of action for how tetrachloroethylene increases MCL, so it was not possible to distinguish whether exposure to tetrachloroethylene results in initiation of new tumors or enhances the ongoing expansion or promotion of existing tumors.

Hepatic Cancer

Evidence for a statistically significant increase in hepatic tumors was observed in male and female mice after oral or inhalation exposure. Like MCL, the biological significance of these increases was debated by the committee because $B6C3F_1$ mice have a high background incidence of hepatic cancer (about 20%). However, the findings were reproduced among several studies and conducted in different laboratories and showed a dose-response relationship. There is also fairly substantial information for characterizing potential modes of action for hepatic tumor formation relative to the data available on MCL and renal cancer.

(See Chapters 6 and 7 and the section below on Mode of Action Analysis.) While the committee recommended that EPA revise its presentation of the mode of action evidence for tetrachloroethylene-related hepatic cancer to clarify its position, the majority of the members agreed with EPA that the mode of action is complex and remains to be established. These members also agreed there was insufficient evidence to rule out human relevance. One member objected to these conclusions and the committee's support of using liver cancer to quantify risk. He concluded that in the absence of evidence of other contributing modes of action, the evidence is sufficient to conclude that the mode of action in mice is predominantly through activation of the peroxisome proliferator activated receptor α, a mode of action he considered to be of little relevance to humans. His arguments are presented in a dissenting statement in Appendix B of the report.

Renal Cancer

Tetrachloroethylene caused a low rate of induction of renal tumors in rats. Although the increases were not statistically significant when compared with concurrent controls, EPA has used historical controls to calculate the chances of two of these rare carcinomas to occur by chance to be less than 0.001. Furthermore, a dose-response trend was shown against the low background and the tumors in the treated rats were malignant whereas the tumors in the controls were not. EPA provided a strong evaluation of the potential modes of action evaluation for tetrachloroethylene-induced kidney cancer. The committee agrees with the agency that mode of action of tetrachloroethylene tumorigenesis is not understood, but that a mutagenic mode of action cannot be ruled out. Thus, kidney tumors observed in tetrachloroethylene-treated rats was considered relevant to humans, even thought the epidemiologic evidence of an association is weak (see Chapter 7).

SELECTION OF TUMOR TYPE FOR QUANTITATIVE ASSESSMENT

The committee was unable to reach consensus on the selection of the critical cancer end point. The majority of members judged that the uncertainties associated with MCL (particularly the high background incidence, uncertainty about the dose response, and poor understanding of mode of action) were too great to support using the data over that of hepatic or renal cancer for determining quantitative estimates of risk. These members judged that the use of the MCL data could only be justified if it is EPA's policy to choose the most conservative unit risk when considering a range of options, but that such justification should be distinguished as a policy decision and not a scientific one. They believe that a more scientifically defensible approach would be to use the data set with the least uncertainty, rather than the data set that yields the most conservative estimate of risk. In their estimation, the hepatic cancer data would have the least uncertainty associated with it, followed by kidney cancer and MCL. The comparison of risk estimates presented in the draft IRIS assessment indicates that a unit risk based on hepatic cancer would be approximately eight-fold less than the estimate based on MCL. A unit risk based on kidney cancer would be five-fold less.

Other members judged that the MCL data should be used for cancer risk estimation. Their opinions were based on the observation that reproducible, statistically significant increases in MCL in male and female rats above the background incidence of MCL were found, and that MCL was the cancer end point with the highest magnitude of response. These members believe that use of the most sensitive response to quantify cancer risk decreases the uncertainty associated with potential differences in metabolism and susceptibility to tetrachloroethylene across exposed populations. They concluded that additional statistical analyses of the dose-response data and the addition of supporting mechanistic information identified by the committee would strengthen existing support for MCL in the draft assessment.

MODE-OF-ACTION ANALYSIS

EPA included mode of action (MOA) analyses for cancer in its draft IRIS assessment (Section 4.4.4, pp. 4-16 to 4-35, for the liver and Section 4.5.4, pp. 4-42 to 4-51, for the kidney). EPA's cancer guidelines present a framework for judging whether available data support a hypothesized MOA of an agent. The application of the framework is best demonstrated in EPA's MOA analysis for renal cancer (see Chapter 7). For hepatic cancer, the committee found that the assessment relies too heavily on generic information on peroxisome proliferators and needs to be focused on tetrachloroethylene and its metabolites.

Chapters 6 and 7 provide more specific guidance on how to improve the presentation of the MOA evidence on tetrachloroethylene-induced hepatic and renal cancer. In general, the committee observes that discussion of MOA[1] analy-

---

[1]There was some disagreement among the committee members on what constitutes "modes of action" and "key events." In Section 4.4.4 of the draft IRIS assessment, EPA discusses several "topics" relevant to the MOA for hepatic toxicity, including metabolism, receptor activation, genotoxic effects, and nongenotoxic effects. EPA's presentation treats those topics as separate MOAs, but metabolism is presented as a key event or a component of multiple modes of action. Some committee members judged that that treatment was appropriate as an introduction to a discussion of multiple modes of action and was consistent with EPA guidelines. Other members judged that although early key events may occur in different pathways, they converge to produce one effect; these members hold the view that there is one MOA for an observed effect for which there are a number of specified key events (early key events may be derived from a series of pathways). Despite those differing viewpoints, all members of the committee agreed that more focused analyses of the available evidence are necessary to support hypothesized MOAs.

ses would be improved by including the proposed temporal sequence of hypothesized tetrachloroethylene-associated key events (possibly as a diagram). Transparency would be improved by presenting the details of experimental results in tabular form, including the chemical (tetrachloroethylene or specific metabolite), species, strain, sex, dose, route and duration of exposure, and experimental outcome or end point. That would allow the reader to follow the evaluation of the relative potency of tetrachloroethylene, or its metabolites, in inducing both key events and tumors and to consider species and strain differences with respect to the events and tumor formation. Other data relevant to the evaluation of hypothesized MOAs should be included. The advantage of such a presentation is that it makes explicit the consideration of the timeline of key events in the context of dose, concordance or lack of concordance between early and late events, and the relative contribution of chemical-specific data compared with generic information on categories of chemicals. That should be done for each hypothesized MOA. Even if the data are insufficient to support hypothesized MOAs, the exercise can be used to identify critical data gaps and to inform the direction of future research.

A general difficulty that the committee encountered in reviewing the MOA analyses is the presentation of conclusions without sufficient supporting evidence or reference to prior discussions of the evidence. Much of the experimental evidence was evaluated in other sections of the draft and presumably formed the basis of statements in the MOA considerations. To make the analyses clear, some reiteration of the evidence is needed in discussions of strength, consistency, and specificity of association of the tumor response with key events; dose-response relationships; temporal associations; and biologic plausibility. Coherence of the database is necessary for characterizing the evidence supporting a MOA. The analysis needs to take into account concordance of dose-response relationships between hypothesized key events and end events and to recognize that key events are necessary but might not be sufficient (in their own right) to induce the adverse outcome.

AGE-DEPENDENT ADJUSTMENT FACTOR

Section 6.2.2.1 of the EPA draft (p. 6-24) states that "age adjustment factors for early life exposures as discussed in the *Supplemental Guidance for Assessing Susceptibility for Early-Life Exposure to Carcinogens* (U.S. EPA 2005b) are not recommended because little evidence exists to indicate that tetrachloroethylene or its oxidative metabolites directly damage DNA, information about genotoxicity of gluthathione (GSH) metabolites in cell assays other than Salmonella or in in vitro experiments are lacking, and the MOA for tetrachloroethylene has not been established." In addition, the assessment reasons that "although a mutagenic MOA would indicate increased early-life susceptibility, there are no data exploring whether there is differential sensitivity to tetrachloroethylene carcinogenicity across life stages." The committee's recommendations for amending sections on genotoxicity and MOA considerations would also strengthen the arguments made by EPA with regard to the need for age-adjustment and low-dose extrapolations. The committee concluded that several metabolites of tetrachloroethylene are clearly genotoxic: *S*-(1,2,2-trichlorovinyl) glutathione (TCVG), *S*-(1,2,2-trichlorovinyl)-L-cysteine, *N*-acetyl-*S*-(1,2,2-trichlorovinyl)-L-cysteine (N-Ac-TCVC), tetrachloroethylene oxide, dichloro-acetic acid (DCA), and chloral hydrate (if it is formed). However, it is questionable whether those metabolites play an important role in the MOA of tetrachloroethylene carcinogenesis in view of their presence in tetrachloroethylene-exposed animals at low or undetectable concentrations and in the absence of convincing evidence of mutagenic and tumor-initiating activity of tetrachloroethylene in vivo. In addition, the committee supports EPA's conclusion that the MOA of tetrachloroethylene is unclear and probably complex. Thus, although the committee agrees that age-dependent adjustment factors for cancer risk should not be applied, given uncertainties with regard to the overall MOA and the biologic relevance of the data on genotoxicity of metabolites of tetrachloroethylene, the rationale for this conclusion should be revisited.

LOW-DOSE EXTRAPOLATION

For cancer risk assessment, EPA relied on the default option of low-dose linear extrapolation to estimate inhalation unit risks and oral slope factors for tetrachloroethylene. EPA describes low-dose linear extrapolation in detail (in Section 5.4.4 of the draft). It entails three steps. First, a dose-response model, typically a mathematical function in the absence of MOA information, that appropriately fits observed data within the experimental data range must be identified. Second, a point of departure (POD) (a benchmark dose or benchmark concentration) along the fitted dose-response model is determined; it corresponds to an exposure that typically induces about 5-10% extra risk above the control's response rate. Then the associated extra cancer risk is divided by the POD to yield a unit risk or a slope factor.

In the draft IRIS assessment, EPA illustrates low-dose extrapolation with six datasets, hepatocellular adenoma or carcinoma in male and female mice (JISA 1993), hemangiosarcoma in male mice (JISA 1993), MCL in male and female rats (JISA 1993), and renal tumors in male rats (NTP 1986). EPA considered multistage models as well as multistage Weibull models for dose-response modeling in conjunction with the dose metric of total metabolism and administered concentration but presented results only of multistage models. It justified the use of the multistage model (p. 5-50) on the basis that MOA information is lacking and that the model has "some parallelism to the multistage carcinogenic process and it fits a broad array of dose-response patterns. Occasionally the multistage model does not fit the available data, in which case an alternate model should be considered." In the case of hepatocellular adenoma and carcinoma in male mice, hemangiosarcoma in male mice, and MCL in female rats, the multistage model does not fit the data at lower doses, as acknowledged by EPA (Figures 5-8a, 5-10a, and 5-12a). EPA did not explain the possible underlying reasons for low-dose nonlinearity and potential adjustment. EPA considered those poor-fit models acceptable solely on the grounds that statistical tests for goodness of fit were not significant ($p > 0.10$). The committee notes that the lack of significance of goodness-of-fit tests can be the result of a small number of animals in each dose group. For example, by doubling the number of animals per dose group while keeping the incidences of tumor the same as in the original dataset of hepatocellular adenoma and carcinoma in male mice (JISA 1993), we can fit the same model (Table 5-11) to the "larger" experiment. The goodness-of-fit test would reach a p value of 0.04, which suggests a poor fit. Alternatively, if we were to fit a multistage model with a (polynomial) degree of 2 to the original data, the goodness-of-fit test would have a p value of 0.25, which would suggest a better fit than the model chosen by EPA (Table 5-11). Thus, using the goodness-of-fit test to justify a selection of a dose-response model can be misleading. Furthermore, contrary to the statement that "dose-response modeling of the candidate data sets presented no particular difficulties" (EPA 2008, p. 5-69), the benchmark dose software automatically fixed some parameters to zero to obtain convergence in model fitting. For example, in the case of hepatocellular adenoma and carcinoma in male mice, the second order coefficient (q_2=0) is fixed but the third order coefficient (term q_3) is not. The criteria under which EPA selected parameters and fixed them was unclear. Also, the parameter q_0 reported in Tables 5-10 and 5-11 should be reported as 1-exp(-q_0) to be consistent with the specification of multistage model in section 5.4.4.1. The committee also notes that the polynomial order used in the multistage dose-response models is limited by the number of dose groups in each experiment; only lower-order multistage models can be fitted, and they are forced to be nearly linear in the low dose range. Therefore, the similarity between the slope of the models and the unit risk taken from the models reflects more on the nearly linear model imposed on the data than the true shape of the dose-response curve. The questionable fitting of a multistage model to some candidate datasets and the insufficient consideration of alternative models in these situations appear to be inconsistent with EPA's cancer-risk guidelines and can contribute to underestimation of the overall uncertainties.

Once a dose-response model was chosen, EPA carried out the estimation of benchmark concentration with its lower confidence limit (BMCL) at a 10% extra risk (5% in one case). The BMCL is used as a POD for unit risk or slope factor. EPA's choice, estimation, and presentation of PODs are adequate and clear.

EPA adopted linear low-dose extrapolation, the default option, with several justifications. First, MOA information is insufficient, and support for dynamic models unavailable. Therefore, nonlinear mechanistic models are unavailable for dose-response modeling. Second, because mathematical models are subject to uncertainties for low-dose extrapolation beyond the experimental dose range, linear extrapolation is more conservative than all sublinear (curvilinear) dose-response models. When individual thresholds in the human population are plausible, wide variation in threshold values implies a curvilinear shape of the dose-response relationship on the average. Thus, linear extrapolation protects susceptible subpopulations (NRC 2009). Third, a few of the candidate datasets, especially EPA's preferred male-rat MCL data, exhibit a linear pattern of dose-response relationships. Whereas those arguments are consistent with EPA's *Guidelines for Carcinogen Risk Assessment*, there is evidence in the candidate datasets that the underlying dose-response relationship can be even supralinear (for example, in female-rat MCL). When that is the case, low-dose linear extrapolation is not conservative. The full range of variation and uncertainty in relation to model choice is not presented, in part because EPA did not consider the possibility of other forms of nonlinear dose-response models, including supralinear, for all candidate datasets.

PHYSIOLOGICALLY BASED PHARMACOKINETIC MODELS, DOSE METRICS, AND INTERSPECIES SCALING

The draft IRIS assessment appears to do a thorough job of reviewing the pertinent scientific literature on the toxicokinetics of tetrachloroethylene. EPA considered several independent efforts to develop physiologically based pharmacokinetic (PBPK) models for tetrachloroethylene and used them to estimate human equivalent doses in terms of environmental exposure and to perform route-to-route extrapolation. In the sections below, the committee reviews EPA's decisions about what PBPK models to use, its choice of dose metrics, and approaches to species extrapolation.

The committee reviewed the original papers describing the selected PBPK models and supporting studies, which in some cases provided the experimental data used to validate model predictions. Evaluation of dose metrics was based on two primary criteria: the ability of the PBPK models to provide discrete estimates of a metric (such as peak blood levels or AUC of the parent chemical or metabolite in blood or target tissue) and the relevance of the parent compound or metabolites to the toxic end point. For cancer, the available evidence suggested that various tetrachloroethylene metabolites were involved or responsible, depending on the end point.

Physiologically Based Pharmacokinetic Modeling Approaches

There have been an unusually large number of independent efforts to develop PBPK models for tetrachloroethylene. EPA is to be commended for its willingness to use the PBPK modeling approach and to explore or test the various published PBPK models for tetrachloroethylene in its risk assessment. EPA used three PBPK models (Rao and Brown 1993; Reitz et al. 1996; Bois et al. 1996). However, there is a notable lack of critical evaluation of the models. Because the most important differences between the models is in prediction of tetrachloroethylene metabolism, there should be more discussion of the pros and cons of using a population-modeling approach as in the Bois et al. (1996) study vs the other models, which rely more directly on animal in vitro and in vivo data. In particular, there seems to be a divergence between the two approaches particularly in estimating the fraction metabolized after smaller tetrachloroethylene exposures. For example, the recent paper by Chiu and Bois (2006) suggests that much higher fractions (23% of the dose) of tetrachloroethylene are metabolized in humans after low exposure (less than 1 ppm).

Reading the descriptions of previous PBPK modeling efforts makes it clear that it would have been preferable for EPA to pursue development of a "harmonized" PBPK model (as was done for trichloroethylene), which synthesized important aspects of the various models (the use of multiple exposure routes and inclusion of all relevant tissue compartments) into a single model. In connection with this recommendation, it is important to recognize that most PBPK models of tetrachloroethylene (and trichloroethylene) are highly derivative of the PBPK model for methylene chloride published by Andersen et al. (1987). The differences between the tretrachloroethylene models are associated with inclusion or exclusion of routes of exposure and the use of experimental data to select parameters for models and validate model predictions. The approach pursued by EPA, using three PBPK models, is a reasonable alternative for the tetrachloroethylene risk assessment for which the goal is to estimate tetrachloroethylene dosimetry related to inhalation exposure. The population pharmacokinetic modeling approach used in the Bois et al. model empirically estimates metabolism parameter values to provide an adequate fit of observed tetrachloroethylene exposure data. Initial estimates (prior distributions) in the Bois et al. model were obtained from the literature by using many sources, and the final estimates (posterior distributions) were obtained by using a Markov-Chain-Monte Carlo approach.

It would have been preferable to use a single PBPK model. All three of the selected models are adequate for characterizing parent-compound (tetrachloroethylene) dosimetry, but they are not equivalent in characterizing tetrachloroethylene metabolism. There is inadequate justification for the selection of dose metrics for tetrachloroethylene metabolism, particularly in the use of total metabolites as the overall dose metric for cancer. The risk assessment would be improved if more effort were devoted to estimating the fraction of an absorbed tetrachloroethylene dose that enters the GSH pathway and the fraction entering the cytochrome P-450 pathway, which leads to the formation of trichloroacetic acid (TCA). That would permit development of more discrete, rational, and defensible dose metrics (for example, total GSH metabolites vs P-450 metabolites) for cancer end points.

The committee recommends that EPA pursue development of a single "harmonized" PBPK model that includes all routes of exposure (inhalation, oral, and dermal) and all relevant tissue compartments. With regard to metabolic dose metrics, the initial goal should be to predict the fraction of an absorbed tetrachloroethylene dose that enters the GSH pathway (initially forming TCVG) and the fraction that enters the P-450 pathway (eventually leading to TCA formation). That would permit development of more discrete dose metrics (such as total GSH metabolites vs P-450 metabolites) and should lead to a more rational and defensible selection of dose metrics for the various cancer end points.

Given the incomplete data available for characterizing the GSH pathway, several approaches may need to be adopted that rely on rodent in vitro data, human in vitro data where available, and allometric scaling as needed. For some key reactions, a parallel approach with trichloroethylene metabolism might be considered; in this respect, the approach and recommendations described by Lash and Parker (2001) should be considered and tested with appropriate model simulations. If modeling the GSH pathway is determined to be infeasible, total metabolism can be used as a reasonably conservative dose metric.

The PBPK model could then be built and tested around a combination of blood tetrachloroethylene and TCA concentrations, in vitro metabolism data, and urinary-excretion data for various metabolites (such as TCA, *N*-Ac-TCVC). With a single harmonized PBPK model, the population modeling approach could be used more effectively to estimate a range of V_{max} and K_m values and compare these posterior distributions with a more robust dataset of blood, in vitro, and urinary-excretion data.

Dose-Metric Selection

The rationale for selection of most dose metrics is clearly explained in the draft IRIS assessment. However, the committee is concerned about the selection of the dose metrics chosen for tetrachloroethylene metabolism. As thoroughly reviewed in the draft, tetrachloroethylene metabolism can be separated into cytochrome P-450-derived oxidative metabolites produced primarily by the liver (the P-450 pathway) and metabolites derived from the initial formation of a GSH conjugate (the GSH pathway) and later reactions in several tissues, including the kidney. The P-450 pathway produces several metabolites, including the biologically persistent metabolite TCA. The P-450 pathway is more closely linked to hepatic cancer in rodent models whereas the GSH pathway appears to be associated more with renal tumors and perhaps leukemia. EPA has chosen not to estimate the flux of metabolism through the GSH pathway and summarizes the rationale for that decision as follows (p. 5-48): "However, the measurements of glutathione-dependent metabolism are from in vitro studies or they are measures of urinary excretion products and are, therefore, not representative of the toxic species in vivo." Instead, the dose metric of total metabolism is used for all cancer end points in which tetrachloroethylene metabolites are implicated. That approach has created several potential problems that are not adequately addressed in the draft. The rationale for excluding the GSH pathway is inconsistent with the use of the three PBPK models, which also use in vitro data (the Reitz model) or urinary-excretion data (the Rao and Brown model) to estimate total metabolism. A fair question to ask is why the use of in vitro data and measures of urinary excretion products was acceptable for the P-450 pathway but not the GSH pathway. The use of total metabolism as a dose metric reflects primarily the P-450 pathway because of large differences between the pathways in the flux of metabolism. The approach used by the different PBPK models to estimate metabolism and specifically estimation of the key metabolic parameters V_{max} and K_m varies substantially. Estimation of total metabolite formation in humans with the Reitz model relies primarily on in vitro hepatic metabolism data (microsomal metabolism, hence only the P-450 pathway) whereas the Rao and Brown model is validated by urinary excretion of nontetrachloroethylene radioactivity and TCA (also reflective primarily of the P-450 pathway). Although there is less experimental information on the GSH pathway, there are in vitro data from two studies that characterize the formation of TCVG in rodents (Dekant et al. 1998; Lash et al. 1998). The Dekant et al. (1998) study also attempted to measure TCVG in human tissues but was unable to detect it. However, their analytic methods appear to be rigorous and to allow estimation of the highest formation rate that could have occurred (still producing undetectable concentrations of TCVG), which would be helpful for risk assessment. A summary of the rates of TCVG formation in the liver in those studies is presented in Table 11-1. These values could be used to estimate the in vivo formation clearance of TCVG in the liver and kidney (data available but not included in Table 11-1) with an approach outlined by Houston and Carlile (1997). It would have been valuable if an attempt had been made to estimate the flux of tetrachloroethylene metabolism through TCVG in rodents and compare it with that in humans by using the results of Dekant et al. (1998) as an upper limit of formation rate. The modeling exercise could be strengthened by integrating the human urinary-excretion data reported by Volkel et al. (1998), for example, on detection of *N*-Ac-TCVC but not DCA in tetrachloroethylene-exposed volunteers.

With respect to hepatic cancer, it is debatable whether it is preferable to use a trichloracetic acid dose metric as opposed to total metabolites. The recent paper by Sweeney et al. (2009) makes a strong argument for the former. However, given the potential role of other P-450 pathway-derived tetrachloroethylene metabolites (discussed in Chapter 6) in hepatic cancer, the use of total metabolites as the dose metric appears justified. In addition, experimental evidence suggests that the toxicity of a directly administered metabolite does not reflect that of the "formed" metabolite (TCA in the case of tetrachloroethylene) even when blood concentrations are comparable (Pang 2009).

The use of total metabolites as a dose metric for renal cancer is not well supported. According to the available data (see Chapter 7), tetrachloroethylene metabolites derived from the GSH pathway are most likely to be the causative agents. Thus, a dose metric that more accurately reflects the flux of metabolism through the GSH pathway would be preferred. For reasons discussed previously, total metabolites constitute essentially a dose metric for the P-450 pathway. The committee encourages EPA to put forth a more thorough effort to develop a TCVG-based dose metric for rodents and possibly humans by using the available data summarized in Table 11-1.

TABLE 11-1 Summary of Data on Hepatic Metabolism of Tetrachloroethylene and Urinary Excretion of Glutathione-Pathway Metabolites

	Species			
Metabolite Formation	Rat (nm/mg per hour)	Mouse (nm/mg per hour)	Human (nm/mg per hour)	Reference or Initial Substrate Concentrations
♂-TCVG (hepatic cytosol)	10.6 ± 2.2	20.7 ± 4.7	nm	Lash et al. (1998)
♂-TCVG (hepatic microsomes)	8.4 ± 1.1	24.5 ± 3.4	nm	2 mM tetrachloroethylene
♀-TCVG (hepatic cytosol)	6.2 ± 0.8	16.3 ± 1.3	nm	5 mM GSH
♀-TCVG (hepatic microsomes)	3.7 ± 0.9	15.6 ± 1.2	nm	
♂-TCVG (hepatic cytosol)	5.1 ± 0.7	1.7 ± 0.4	<0.06	Dekant et al. 1998
♂-TCVG (hepatic microsomes)	nd	nd	<0.06	3 mM tetrachloroethylene
♀-TCVG (hepatic cytosol)	1.2 ± 0.5	1.6 ± 0.4	<0.06	5 mM GSH
♀-TCVG (hepatic microsomes)	nd	nd	<0.06	
Cumulative urinary excretion	nm/kg		nm/kg	
N-Ac-TCVC (both sexes)	3.5 – 415		0.65 – 3.01	Volkel et al. 1998
TCA	1.9 – 66.7		70 – 290	Tetrachloroethylene in vivo
DCA	0.25 – 5.5		nd	10-40 ppm

Note: TCA data included for comparison.
Abbreviations: DCA = dichloroacetic acid; GSH = glutathione; *N*-Ac-TCVC = *N*-acetyl-*S*-(1,2,2-trichlorovinyl)-L-cysteine; nd = not detected; nm = not measured; TCA = trichloroacetic acid; TCVG = *S*-1,2,2-trichlorovinyl glutathione.

Interspecies Scaling

The approach used for interspecies scaling is presented in a reasonably clear manner. Figure 5-7 of the draft and the discussion on pp. 5-53 to 5-55 are particularly helpful. The committee's main concern in this regard is how errors in estimating the metabolized fraction affect the extrapolation process.

Extrapolation from Route to Route

EPA has chosen to use the venous-blood area under the curve (AUC) as the route-to-route dose metric for extrapolating an inhalation exposure to a corresponding oral dose. The rationale for this approach is sound and adequately explained in the draft document. However, its implementation raises serious methodologic concerns based on inappropriate use of the selected PBPK models and uncertainties in the fraction of an oral dose of tetrachloroethylene that is metabolized. The three PBPK models used by EPA were specifically formulated and validated against inhalation exposures. There was no attempt to validate model predictions against blood tetrachloroethylene concentrations after oral dosing. To use the PBPK models, EPA has empirically assumed a value of the rate of oral absorption of tetrachloroethylene, which is entered as a constant. That approach is inferior to direct estimation as used in other published PBPK models, such those by Gearhart et al. (1993) and Dallas et al. (1995) (the latter only for rats and dogs). These PBPK models would have been better choices to begin the extrapolation exercise. Better still, a harmonized PBPK modeling approach (recommended earlier in this chapter) would have provided the greatest confidence in the route-to-route extrapolation.

Aside from the use of an appropriate PBPK model (for example, one specifically formulated and validated against oral-dosing data), uncertainty is associated with the dose dependence of tetrachloroethylene metabolism. EPA has assumed that a person will have nine drinking-water events during a day at roughly 2-hour intervals (excluding nighttime). The calculated oral equivalent dose of tetrachloroethylene is 1.1 mg/kg per day or 0.122 mg/kg per dose (that is, the discrete tetrachloroethylene dose received in each drinking-water episode). That oral dose is an order of magnitude lower than those previously used in toxicokinetic studies of tetrachloroethylene. The data from past studies clearly suggest that the fraction of a tetrachloroethylene oral dose that is metabolized is progressively reduced as the dose increases (Pegg et al. 1979; Frantz and Watanabe 1983; Schumann et al. 1980; Dallas et al. 1995). The issue of uncertainty in fractional tetrachloroethylene metabolism and dose was also raised by Reitz et al. (1996), whose PBPK model was used by EPA for route-to-route extrapolation of total metabolites. That raises the serious concern that a much greater fraction of the 0.122-mg/kg dose of tetrachloroethylene is being metabolized than was predicted by the PBPK models used in the risk assessment. The impact of the probable error is that the estimates of the venous-blood AUC of tetrachloroethylene shown in Figure 5-3 of the draft are probably overpredicted (that is, a higher oral dose is needed) and the estimates of total metabolites are underpredicted and may affect cancer assessments.

UNCERTAINTY

Cancer risk assessment results in an overarching summary of cancer risk by using a unit risk or a slope factor. In the process of deriving the unit risk or slope factor, uncertainty at every step is propagated into the final estimate. Because of the quantitative nature of the final risk estimates, it is critical to understand the effects of uncertainties on risk estimates both qualitatively and quantitatively. EPA has clearly identified key sources of variation and uncertainty in the process of risk assessment, including human population variation (susceptibility in exposure, metabolism, and response to exposure), low-dose extrapolation (including choice of dose-response models), choice of dose metric, extrapolation from animals to humans (cross-species scaling), and the use of PBPK models for route-to-route extrapolation. EPA's investigation of the effects of uncertainties on risk estimates is qualitative except in dealing with such issues as the choice of dose-response models, the use of PBPK models, and, to a small degree, variation between studies. The following is an appraisal of EPA's uncertainty analysis.

EPA's presentation of the uncertainty analysis is generally transparent and includes sufficient detail. The tabular presentation of choices of study, end points, the approach (models) to extrapolation, and their effects on risk estimates is especially informative and easy to follow. For example, Table 6-3 of the draft summarizes key characteristics of the candidate rodent experiments and associated tumor types. Whereas that form of presentation is helpful, the committee does not agree with all characterizations presented in the table (see earlier discussion about the different cancer end points).

Similarly, Table 6-5 highlights EPA's choices and their effects on the determination of the upper bound of the risk estimate at many critical steps of the risk-estimation process. It also lists EPA's decision and the corresponding justification. Such presentation is effective and should be fully used. In some instances, however, the justification of EPA's choice is debatable. As discussed in Chapter 6, for example, the committee believes that the hepatic-tumor data on male and female mice should also be given strong weight for consideration on the basis of dose-response data. In the case of the choice of dose-response model from among the multistage, Weibull, log-probit, and log-logistic models, one justification for using a multistage model was the relatively small variation in unit risk among the four models (a factor of 1.4). However, that narrow variation was shown only in the male-rat MCL data. The MCL data exhibit a nearly linear dose-response relationship and hence attenuate the difference among the four models. If EPA would consider other bioassay or tumor sites (such as hepatic tumors in female mice or MCL in female rats) that show a somewhat more nonlinear shape of the dose-response relationship, the variation in unit risk calculated by the models would be much greater. Even in the case of MCL in male rats, the risk obtained by linear extrapolation to 1.5 × 10-5 mEq/kg per day varied by up to several orders of magnitude among the same four models (Table 5B-2). Therefore, choosing a multistage model on the basis that risks with other models at a POD are similar is difficult to justify.

More detail would have been helpful in a few of EPA's analyses of uncertainties. For example, EPA's assessment of uncertainties under different model forms (multistage, Weibull, log-probit, and log-logistic) used bootstrap simulations. The results show variation in extra risk spanning orders of magnitude at the low dose of 1.5×10^{-5} mEq/kg per day (bootstrap mean, 9.172×10^{-7} to about 1.078×10^{-3} in Table 5B-2) among the models despite their comparable goodness of fit to the dataset on MCL in male rats. Details about the bootstrap methods and scheme would facilitate appropriate understanding of the bootstrap distributions. For example, what was the number of bootstrap replications? What bootstrap method was used to simulate the distribution of extra risk? The committee views EPA's consideration of uncertainty due to different forms of the dose-response relationship highly valuable, and it encourages EPA to extend such quantitative evaluation to all candidate datasets so that a fuller array of uncertainties can be assessed.

The committee notes that EPA discusses uncertainties in detail. However, the discussion typically focuses on individual sources without an in-depth illustration of the propagation of the uncertainties and their cumulative effect on the final risk estimate. That limitation is in part the result of qualitative treatment of uncertainties in many instances, notably concerning MOA, the choice of bioassay, and human variations. New methods that allow probabilistic quantification of the overarching uncertainty and of the variation in the final risk estimate are emerging (see Chapter 12). The capability to quantify the full range of overarching uncertainties associated with risk estimates facilitates separation of the science of risk assessment from risk-management decision-making. The committee encourages EPA to consider recommendations in *Science and Decisions* (NRC 2009) regarding uncertainty and variability.

12

Moving Beyond the Current State of Practice

The committee found that the Environmental Protection Agency (EPA) draft Integrated Risk Information System (IRIS) assessment could be improved in several ways. Such changes are not necessary for completing the current assessment but should be considered when tetrachloroethylene is re-evaluated. They include improvements in the presentation and organization of information, addition of transparency in documenting procedures used for identifying and selecting studies, and the use of evolving approaches to uncertainty analysis. Guidance in many of these areas is provided in a recent National Research Council (NRC 2009) report *Science and Decisions*, which discusses advancing risk assessment practices.

ORGANIZATION AND APPROACH

There is a vast amount of literature on tetrachloroethylene, and drafting of the IRIS assessment was hampered by the need to manage such a large volume. EPA should consider ways of reorganizing the document to streamline the presentation of data and analyses. The current organization requires that some information be duplicated in various places. Part of the document also appears to be targeted to controversies in interpretation of some aspects of the data. In several instances, the committee found that EPA had spent more time in debunking others' positions than in bolstering its own arguments.

Although the draft provides a comprehensive review of the available data, it is not clear whether studies were evaluated case by case or a consistent set of criteria were applied. To ensure consistent and transparent analysis of the data, criteria for identifying, analyzing, and selecting studies should be established in advance to guide the assessment in focusing on the most relevant studies. Study design and methods are the most important factors in study selection. Other factors, such as exposure considerations and outcomes, will also play a role in selection.

Consideration of the quality of an assessment is predicated on not only its content but the process by which it was prepared. There should be a preassessment discussion of problem formulation and issue identification that indicates the extent of reliance on previous reviews, the focus of the future effort, and the specific issues on which the assessment is likely to be focused. (Guidance on the design of a risk assessment in its formative stages is provided by the NRC [2009].) That would serve as a basis for soliciting external multidisciplinary input at an early stage in such critical matters as mode of action and evaluation of information on specific end points (including both toxicologic and epidemiologic data). It would include a priori delineation and weighting of criteria for evidence of hazard and options analysis for dose-response assessment and associated uncertainties. Attention to specifying evaluation criteria and the options considered is expected to contribute considerably to transparency in the separation of science judgment from science-policy choices.

To increase transparency, accountability, and defensibility and to improve the content and process of assessments, the committee offers the following recommendations regarding future assessments of tetrachloroethylene:

- The nature of, timeframe for, and extent of consideration of relevant data should be clearly framed and stated (for example, standard searching of identified electronic sources with criteria specified, cutoff date past which no additional data were considered, and identification of current studies by reviewers).
- Exclusion criteria for particular studies should be clearly identified and explained (for example, unpublished or published after a particular date). In particular, there should be description of steps taken to ensure that studies identified after the original search were selected without bias from the totality of the available data.
- The methods used for qualitative characterization of uncertainties should be clearly identified, explained, and documented. Qualitative assessment of uncertainty involves (WHO 2008) evaluation of the level of uncertainty of each specified source according to a scoring method, identification and description of the major sources of uncertainty, appraisal of the knowledge base associated with each major source of uncertainty, identification of controversial sources of uncertainty, evaluation of the subjectivity of choices of controversial sources of information, and iteration until the output reflects the current state of knowledge.
- The specific nature of the process of preparing and reviewing the assessment—including identification of authors and reviewers, timeline and nature of peer input, consultation, and peer review—should be set forth.

UNCERTAINTY ASSESSMENT

Scientific Needs

Beginning as early as the 1980s (NRC 1983), expert scientific advisory groups have been recommending that risk analyses include a clear discussion of the uncertainties in risk estimation. The National Research Council (NRC 1994; 2009) stated the need to describe uncertainty and to capture variability in risk estimates. The Presidential/Congressional Commission on Risk Assessment and Risk Management (PCCRARM 1997) recommended against a requirement or need for a "bright-line," or single-number, level of risk. Regulatory science often requires selection of a limit for a contaminant, but the limit always contains uncertainty as to how protective it is. Explicit quantification of uncertainty enables decisions regarding degree of protection to be made in the policy arena rather than buried among assumptions of a technical analysis. Risk characterization became EPA policy in 1995, and the principles of transparency, clarity, consistency, and reasonableness are explicated in the 2000 *Risk Characterization Handbook* (EPA 2000). Criteria for transparency, clarity, consistency, and reasonableness require analysts to describe and explain the uncertainties, variability, and known data gaps in a risk analysis and imply that decision-makers should explain how they affect resulting decision-making processes (EPA 1992, 1995, 2000).

On numerous occasions, the National Research Council has explicitly called for the use of probabilistic risk assessment (NRC 2006b, 2007). In 1983, it formalized the risk-assessment paradigm that includes dose-response analysis as a key component (NRC 1983). In 1989, it recommended that EPA consider the distribution of exposure and sensitivity of response in the population (NRC 1989). In 1991, it stated that when assessing human exposure to air pollutants, EPA should present model results with estimated uncertainties. In 1993, it recommended that EPA thoroughly discuss uncertainty and variability in the context of ecologic risk assessment (NRC 1993). In 1994, in a major review of risk-assessment methodology, it stated that "uncertainty analysis is the only way to combat the 'false sense of certainty,' which is *caused* by a refusal to acknowledge and (attempt to) quantify the uncertainty in risk predictions" (NRC 1994). And in 2002, it suggested that EPA's estimation of health benefits was not wholly credible, because the agency failed to deal formally with uncertainties in its analyses (NRC 2002).

EPA's Science Advisory Board (SAB) has made recommendations similar to those of the National Research Council. It urged EPA to characterize variability and uncertainty more fully and more systematically and to replace single-point uncertainty factors with a set of distributions by using probabilistic methods (EPASAB 2007). EPA has developed numerous internal handbooks on conducting quantitative analysis of uncertainties in various contexts (e.g., EPA

1995, 1997, 1998, 2000, 2001). In 2009, it provided a detailed overview of the current use of probabilistic risk analysis in the agency (including 16 detailed case-study examples), an enumeration of the relevance of probabilistic risk analysis to decision-making, common challenges faced by decision-makers, an overview of probabilistic risk-analysis methodology, and recommendations on how probabilistic risk analysis can support regulatory decision-making. EPA's National Exposure Research Laboratory has recently explored methodologic issues in dealing with uncertainty quantitatively when air-quality, exposure, and dose models are coupled (Ozkaynak et al. 2008).

There are numerous texts on analysis of uncertainty (e.g., Morgan and Henrion 1990; Cullen and Frey 1999; Vose 2008). The World Health Organization (WHO) has recently released guidance on qualitative and quantitative methods of uncertainty analysis in the context of exposure assessment (WHO 2008). Its guidelines have been used by EPA to support uncertainty assessments related to exposure to and health effects of criteria pollutants under the National Ambient Air Quality Standards. Hence, the framework is a general one. In particular, WHO proposed guiding principles that are adapted as follows:

- Uncertainty analysis should be an integral part of the assessment.
- The objective and level of detail of the uncertainty analysis should be based on a tiered approach and be consistent with the overall scope and purpose of the assessment.
- Sources of uncertainty and variability should be systematically identified.
- The presence or absence of moderate to strong dependence of one input on another should be discussed and appropriately accounted for.
- Data, expert judgment, or both should be used to inform the specification of uncertainties in scenarios, models, and inputs.
- Sensitivity analysis should be an integral component of the assessment.
- Uncertainty analyses should be fully and systematically documented in a transparent manner, including quantitative aspects pertaining to data, methods, inputs, models, and outputs; sensitivity analysis; qualitative aspects; and interpretation of results.
- The results of the assessment, including uncertainty, should be subject to an evaluation process that may include peer review, model comparison, quality assurance, or comparison with relevant data or independent observations.
- Where appropriate for an assessment objective, assessments should be iteratively refined to incorporate new data and methods to reduce uncertainty and to improve the characterization of variability.
- Communication of assessment uncertainties to stakeholders should reflect the needs of different audiences in a transparent and understandable manner.

Decision-Making Context for Use of Uncertainty Assessment

EPA decision-makers face scientifically complex problems that entail uncertainty. A risk assessment includes exposure assessment, dose-response assessment, and risk characterization. Methods for quantifying uncertainty in exposure assessment are well accepted and widely applied (e.g., Cullen and Frey 1999). Risk can be characterized for a population (for example, the expected number of excess cancers) or an individual (for example, the incremental lifetime risk of excess cancer). The need for characterization of uncertainty in risk characterization is supported by numerous National Research Council studies (for example, NRC 1994). The decision context of risk assessment includes setting priorities for the activities of the assessment and development of data for the assessment to characterize and, where possible, reduce uncertainty and managing risk. Decision-makers often want to know who is at risk, the magnitude of risk, and tradeoffs between risk-management alternatives. Examples of specific questions that decision-makers may ask include the following (Bloom et al. 1993; Krupnick et al. 2006):

- How representative is the estimate (for example, what is the variability around an estimate)?
- What are the major gaps in knowledge, and what major assumptions are used in the assessment? How reasonable are the assumptions?
- Is it likely that additional data collection and research would lead to a different decision? How long would it take to collect the information, how much would it cost, and would the resulting decision be substantially different?

Moving Beyond the Current State of Practice

EPA's assessment of tetrachloroethylene follows a traditional approach for developing "cancer slope factors" and "hazard indexes" that take into account uncertainties qualitatively and through uncertainty factors. Although EPA claims to have introduced a new method for uncertainty analysis in the context of the dose-response assessment of tetrachloroethylene, in fact the only differences between the draft IRIS aassessment for tetrachloroethylene and those of other chemicals are the consideration of multiple end points and the limited use of bootstrap simulation for only a portion of uncertainties. The various alternative dose-response estimates developed represent inter-end-point variability, not uncertainty.

The well-accepted default-based approach to developing dose-response relationship estimates leads to point estimates, not distributional ranges. The choice of point estimates is based on default assumptions regarding uncertainty factors and default inference methods for fitting and interpretation of dose-response functions. Therefore, such estimates do not depict uncertainty quantitatively in conjunction with the final result, and they are based on assumptions that may mix policy judgments about degree of protection and scientific goals of developing a best estimate. Thus, the state of practice does not fully meet the spirit of principles, guidelines, and recommendations that have accrued over the years from such science advisory bodies as the EPA's SAB, WHO, and most recently the National Research Council (NRC 2009). Today, the approach that EPA has taken is considered to be the best practice but not a state-of-the-art practice. For example, although uncertainty factors are used to account for such issues as extrapolation from subchronic to chronic exposure, interspecies extrapolation, and adjustments from lowest-observed-adverse-effect levels to no observed-adverse-effect levels, the use of such factors does not characterize uncertainty. There is a lack of transparency as to the basis of those factors and whether they mix policy-based assumptions with science-based assessments. Furthermore, a user of the resulting dose-response estimates has no information regarding the quantitative range of uncertainty.

Others have illustrated methods that could be used to quantify uncertainty in dose-response assessment, but such techniques are not reviewed, considered, or applied in EPA's draft assessment of tetrachloroethylene. We mention a few illustrative examples of techniques that others have explored. Evans et al. (1994) demonstrated a probability-tree method for quantifying uncertainty associated with low-dose cancer risk. IEc (2006) has demonstrated a method for quantifying uncertainty in concentration-response functions for fine particulate matter that is based on a formal, systematic approach to eliciting subjective probability distributions from multiple carefully selected experts. Small (2008) enumerates an approach that, if implemented, would advance the state of practice in combining multiple sources of uncertainty, including combination based on judgment and data. In this approach, a prior distribution is postulated to the options on a key assumption, such as the one for MOA, or a key choice, such as candidate data sets. Each final risk estimate is a result of a combined set of assumptions and choices propagating through the risk-assessment process tree and is assigned a probability that results from the prior probabilities assigned to each associated assumption and choice. The collection of all final risk estimates will thus cover all admissible combinations of assumptions and choices and will form a probabilistic distribution that quantifies the full range of variation of the risk estimates. Additionally, this probabilistic distribution of risk estimates can be used, with the incorporation of new data, to obtain posterior probabilities for the assumptions and choices involved in each step of risk estimation. With the help of a distribution of risk estimates to reflect the overarching uncertainties and variations, regulatory policy can be less dependent on a principal study or a few data sets. In fact, the risk-management process can use the distributional properties to choose and justify a final risk estimate in the context of this full range of uncertainties and variations.

Hence, EPA in general and the IRIS program in particular should explore methods for adoption or adaptation to improve the qualitative and quantitative characterization of uncertainty. In general, there should be both well-structured qualitative assessment of uncertainties and quantitative assessments wherever possible. Preference should be given to quantitative assessment as the desirable approach, and justification for the use of qualitative instead of quantitative approaches should be provided. For example, it should be explained why the state of science is adequate to characterize a point estimate but not a range of uncertainty if quantitative methods of uncertainty analysis are not used.

A key way forward in quantifying uncertainty is to accept the role of expert scientific judgment. Such judgment is used routinely to make inferences regarding hazard identification and in developing dose-response characterizations of chemicals. The examples of Evans et al. (1994), IEC (2006), and Small (2008) rely on encoding expert judgment as subjective probability distributions to various degrees. The appropriate selection and application of methods for quantifying uncertainty in dose-response relationships are undergoing development and need additional research from which guidance on best practices can be derived. As an example of the exploratory nature of dealing with uncertainty in dose-response relationships, the 2007 Resources for the Future workshop "Uncertainty Modeling in Dose Response: Dealing with Simple Bioassay Data, and Where Do We Go from Here?" explored a variety of methods for quantifying uncertainty and the needed role of qualitative assessment to deal with aspects of dose-response modeling that are believed not to be amenable to quantification. Some quantitative techniques that were explored were bootstrap simulation and probabilistic inversion with isotonic regression and Bayesian-model averaging to deal with uncertainty in model structure. However, although there is not yet a default method for quantifying uncertainty in dose-response relationships, EPA can and should review and adopt or adapt various methods that are being explored in the scientific community, taking particular note of the possibilities for combining expert judgment and data with Bayesian approaches.

References

Altmann, L., A. Bottger, and H. Wiegand. 1990. Neurophysiological and psychophysical measurements reveal effects of acute low-level organic solvent exposure in humans. Int. Arch. Occup. Environ. Health 62(7):493-499.

Altmann, L., H. Wiegand, A. Bottger, F. Elstermeier, and G. Winneke. 1992. Neurobehavioral and neurophysiological outcomes of acute repeated perchloroethylene exposure. Appl. Psychol. 41(3):269-279.

Altmann, L., H.F. Neuhann, U. Kramer, J. Witten, and E. Jermann. 1995. Neurobehavioral and neurophysiological outcome of chronic low-level tetrachloroethene exposure measured in neighborhoods of dry cleaning shops. Environ. Res. 69(2):83-89.

Anders, M.W., L. Lash, W. Dekant, A.A. Elfarra, and D.R. Dohn. 1988. Biosynthesis and biotransformation of glutathione S-conjugates to toxic metabolites. Crit. Rev. Toxicol. 18(4):311-341.

Andersen, M.E., H.J. Clewell, III, M.L. Gargas, F.A. Smith, and R.H. Reitz. 1987. Physiologically based pharmacokinetics and the risk assessment process for methylene chloride. Toxicol. Appl. Pharmacol. 87(2):185-205.

Anttila, A., E. Pukkala, M. Sallmen, S. Hernberg, and K. Hemminki. 1995. Cancer incidence among Finnish workers exposed to halogenated hydrocarbons. J. Occup. Environ. Med. 37(7):797-806.

Aschengrau, A., D. Ozonoff, C. Paulu, P. Coogan, R. Vezina, T. Heeren, and Y. Zhang. 1993. Cancer risk and tetrachloroethylene-contaminated drinking water in Massachusetts. Arch. Environ. Health 48(5):284-292.

ATSDR (Agency for Toxic Substances and Disease Registry). 1997. Toxicological Profile for Trichloroethylene (Update). U.S. Department of Health and Human Services, Public Health Service, Agency for Toxic Substances and Disease Registry, Atlanta, GA. September 1997 [online]. Available: http://www.atsdr.cdc.gov/tox profiles/tp19-p.pdf [accessed Oct. 13, 2009].

Austin, E.W., J.M. Parrish, D.H. Kinder, and R.J. Bull. 1996. Lipid peroxidation and formation of 8-hydroxydeoxyguanosine from acute doses of halogenated acetic acids. Fundam. Appl. Toxicol. 31(1):77-82.

Beliles, R.P., D.J. Brusick, and F.J. Mecler. 1980. Teratogenic-Mutagenic Risk of Workplace Contaminants: Trichloroethylene, Perchloroethylene and Carbon Disulfide. NTIS Publication No. PB-82 185-075. NIOSH Contract Report No. 201-77-0047. Prepared by Litton Bionetics, Inc., for U.S. Department of Health and Human Services, National Institute for Occupational Safety and Health, Cincinnati, OH.

Benignus, V.A., P.J. Bushnell, W.K. Boyes, C. Eklund, and E.M. Kenyon. 2009. Neurobehavioral effects of acute exposure to four solvents: Meta-analyses. Toxicol. Sci. 109(2):296-305.

Bhattacharya, R.K., and M.O. Schultze. 1971. Properties of DNA isolated from tissues of calves treated with S-(1,2-dichlorovinyl)-L-cysteine. II. Primer-template activity for bacterial DNA polymerases. Arch. Biochem. Biophys. 145(2):575-582.

Bhattacharya, R.K., and M.O. Schultze. 1972. Properties of DNA treated with S-(1,2-dichlorvinyl)-L-cysteine and a lyase. Arch. Biochem. Biophys. 153(1):105-115.

Birner, G., A. Rutkowska, and W. Dekant. 1996. N-acetyl-S-(1,2,2-trichlorovinyl)-L-cysteine and 2,2,2-trichloroethanol: Two novel metabolites of tetrachloroethene in humans after occupational exposure. Drug Metab. Dispos. 24(1):41-48.

Birner, G., U. Bernauer, M. Werner, and W. Dekant. 1997. Biotransformation, excretion and nephrotoxicity of haloalkane-derived cysteine S-conjugtes. Arch. Toxicol. 72(1):1-8.

Blair, A., S.A. Petralia, and P.A. Stewart. 2003. Extended mortality follow-up of a cohort of dry cleaners. Ann. Epidemiol. 13(1):50-56.

Bloom, D.L., D.M. Byrne, and J.M. Andreson. 1993. Communicating Risk to Senior EPA Policy-Makers: A Focus Group Study. Prepared by Bloom Research and the Office of Air Quality Planning and Standards. U.S. Environmental Protection Agency, Research Triangle Park, NC.

Boice, J.D., Jr., D.E. Marano, J.P. Fryzek, C.J. Sadler, and J.K. McLaughlin. 1999. Mortality among aircraft manufacturing workers. Occup. Environ. Med. 56(9):581-597.

Bois, F.Y., A. Gelman, J. Jiang, D.R. Maszle, L. Zeise, and G. Alexeef. 1996. Population toxicokinetics of tetrachloroethylene. Arch. Toxicol. 70(6):347-355.

Bolt, H.M., R.J. Laib, and J.G. Filser. 1982. Reactive metabolites and carcinogenicity of halogenated ethylenes. Biochem. Pharmacol. 31(1):1-4.

Boyes, W.K., M. Bercegeay, W.M. Oshiro, Q.T. Krntz, E.M. Kenyon, P.J. Bushnell, and V.A. Benignus. 2009. Acute perchloroethylene exposure alters rat visual-evoked potentials in relation to brain concentrations. Toxicol. Sci. 108(1):159-172.

Briving, C., I. Jacobson, A. Hamberger, P. Kjellstrand, K.G. Haglid, and L.E. Rosengren. 1986. Chronic effects of perchloroethylene and trichloroethylene on the gerbil brain amino acids and glutathione. Neurotoxicology 7(1):101-108.

Brodkin, C.A., W. Daniell, H. Checkoway, D. Echeverria, J. Johnson, K. Wang, R. Sohaey, D. Green, C. Redlich, D. Gretch, and L. Rosenstock. 1995. Hepatic ultrasonic changes in workers exposed to perchloroethylene. Occup. Environ. Med. 52(10):679-685.

Bull, R.J., I.M. Sanchez, M.A. Nelson, J.L. Larson, and A.L. Lansing. 1990. Liver tumor induction in B6C3F1 mice by dichloroacetate and trichloroacetate. Toxicology 63(3):341-359.

Bull, R.J., G.A. Orner, R.S. Cheng, L. Stillwell, A.J. Stauber, L.B. Sasser, M.K. Lingohr, and B.D. Thrall. 2002. Contribution of dichloroacetate and trichloroacetate to liver tumor induction in mice by trichloroethylene. Toxicol. Appl. Pharmacol. 182(1): 55-65.

Bull, R.J., L.B. Sasser, and X.C. Lei. 2004. Interactions in the tumor-promoting activity of carbon tetrachloride, trichloroacetate, and dichloroacetate in the liver of male B6C3F1 mice. Toxicology 199(2-3):169-183.

Bushnell, P.J., and K.M. Crofton. 1999. Organic solvents. Pp 394-428 in Introduction to Neurobehavioral Toxicology: Food and Environment, R.J.M. Niesink, R.M.A.

Jaspers, L.M.W. Kornet, J.M. van Ree, and H.A. Tilson, eds. Boca Raton, FL: CRC Press.

Bushnell, P.J., W.K. Boyes, T.J. Shafer, A.S. Bale, and V.A. Benignus. 2007. Approaches to extrapolating animal toxicity data on organic solvents to public health. Neurotoxicology 28(2):221-226.

Bushinsky, D.A., F.L. Coe, and O.W. Moe. 2008. Nephrolithiasis. Pp. 1299-1349 in Brenner and Rector's The Kidney, 8th Ed. Philadephia, PA: Saunders.

Caldwell, D.J. 1999. Review of mononuclear cell leukemia in F-344 rat bioassays and its significance to human cancer risk: a case study using alkyl phthalates. Regul. Toxicol. Pharmacol. 30(1):45-53.

Carney, E.W., B.A. Thorsrud, P.H. Dugard, and C.L. Zablotny. 2006. Developmental toxicity studies in Crl:CD (SD) rats following inhalation exposure to trichloroethylene and perchloroethylene. Birth Defects Res. B Dev. Reprod. Toxicol. 77(5):405-412.

Cavalleri, A., F. Gobba, M. Paltrinieri, G. Fantuzzi, E. Righi, and G. Aggazzotti. 1994. Perchloroethylene exposure can induce colour vision loss. Neurosci. Lett. 179(1-2):162-166.

Cederberg, H., J. Henriksson, and M.L. Binderup. In press. DNA damage detected by the alkaline comet assay in the liver of mice after oral administration of tetrachloroethylene. Mutagenesis.

Chiu, W.A., and F.Y. Bois. 2006. Revisiting the population toxicokinetics of tetrachloroethylene. Arch. Toxicol. 80(6):382-385.

Chiu, W.A., S. Micallef, A.C. Monster, and F.Y. Bois. 2007. Toxicokinetics of inhaled trichloroethylene and tetrachloroethylene in humans at 1 ppm: Empirical results and comparisons with previous studies. Toxicol. Sci. 95(1):23-36.

Costa, A.K., and K.M. Ivanetich. 1984. Chlorinated ethylenes: Their metabolism and effect on DNA repair in rat hepatocytes. Carcinogenesis 5(12):1629-1636.

Costa, L.G.., M. Aschner, A. Vitalone, T. Syversen, and O.P. Soldin. 2004. Developmental neuropathology of environmental agents. Annu. Rev. Pharmacol. Toxicol. 44:87-110.

Costantini, A.S., L. Miligi, D. Kriebel, V. Ramazzotti, S. Rodella, E. Scarpi, E. Stagnaro, R. Tumino, A. Fontana, G. Masala, C. Vigano, C. Vindigni, P. Crosignani, A. Benvenuti, and P. Vineis. 2001. A multicenter case-control study in Italy on hematolymphpoietic neoplasms and occupation. Epidemiology 12(1):78-87.

Cullen, A.C., and H.C. Frey. 1999. Probabilistic Exposure Assessment: A Handbook for Dealing with Variability and Uncertainty in Models and Inputs. New York: Plenum Press.

Dallas, C.E., X.M. Chen, K. O'Barr, S. Muralidhara, P. Varkonyi, and J.V. Bruckner. 1994. Development of a physiologically based pharmacokinetic model for perchloroethylene using tissue concentration-time data. Toxicol. Appl. Pharmacol. 128(1):50-59.

Dallas, C.E., X.M. Chen, S. Muralidhara, P. Varkonyi, R.L. Tackett, and J.V. Bruckner. 1995. Physiologically based pharmacokinetic model useful in prediction of the influence of species, dose, and exposure route on perchoroethylene pharmacokinetics. J. Toxicol. Environ. Health 44(3):301-317.

Dekant, W., M. Metzler, and D. Henschler. 1986a. Identification of S-1,2,2-trichlorovinyl-N-acetylcysteine as a urinary metabolite of tetrachloroethylene: Bioactivation through glutathione conjugation as a possible explanation of its nephrocarcinogenicity. J. Biochem. Toxicol. 1(2):57-72.

Dekant, W., M. Metzler, and D. Henschler. 1986b. Identification of S-1,2-dichlorovinyl-N-acetyl-cysteine as a urinary metabolite of trichloroethylene: A possible explanation for its nephrocarcinogenicity in male rats. Biochem. Pharmacol. 35(15):2455-2458.

Dekant, W., S. Vamvakas, H. Berthold, S. Schmidt, D. Wild, and D. Henschler. 1986c. Bacterial beta-lyase mediated cleavage and mutagenicity of cysteine conjugates derived from the nephrocarcinogenic alkenes trichloroethylene, tetrachloroethylene and hexachlorobutadiene. Chem. Biol. Interact. 60(1):31-45.

Dekant, W., K. Berthold, S. Vamvakas, D. Henschler, and M.W. Anders. 1988. Thioacylating intermediates as metabolites of S-(1,2-dichlorovinyl)-L-cysteine and S-(1,2,2-trichlorovinyl)-L-cysteine formed by cysteine conjugate beta-lyase. Chem. Res. Toxicol. 1(3):175-178.

Dekant, W., G. Birner, M. Werner, and J. Parker. 1998. Glutathione conjugation of perchloroethene in subcellular fractions from rodent and human liver and kidney. Chem. Biol. Interact. 116(1-2):31-43.

DeMarini, D.M., E. Perry, and M.L. Shelton. 1994. Dichloroacetic acid and related compounds: Induction of prophage in E. coli and mutagenicity and mutation spectra in Salmonella TA100. Mutagenesis 9(5):429-437.

Doherty, A.T., S. Ellard, E.M. Parry, and J.M. Parry. 1996. An investigation into the activation and deactivation of chlorinated hydrocarbons to genotoxins in metabolically competent human cells. Mutagenesis 11(3):247-274.

Dosemeci, M., P. Cocco, and W.H. Chow. 1999. Gender differences in risk of renal cell carcinoma and occupational exposures to chlorinated aliphatic hydrocarbons. Am. J. Ind. Med. 36(1):54-59.

Doyle, P., E. Roman, V. Beral, and M. Brookes. 1997. Spontaneous abortion in dry cleaning workers potentially exposed to perchloroethylene. Occup. Environ. Med. 54(12):848-853.

Dreessen, B., G. Westphal, J. Bunger, E. Hallier, and M. Muller. 2003. Mutagenicity of the glutathione and cysteine S-conjugates of the haloalkenes 1,1,2-trichloro-3,3,3-trifluoro-1-propene and trichlorofluoroethene in the Ames test in comparison with the tetrachloroethene-analogues. Mutat. Res. 539(1-2):157-166.

Echeverria, D., R.F. White, and C. Sampaio. 1995. A behavioral evaluation of PCE exposure in patients and dry cleaners: A possible relationship between clinical and preclinical effects. J. Occup. Environ. Med. 37(6):667-680.

Elfarra, A.A., and R.J. Krause. 2007. S-(1,2,2-trichlorovinyl)-L-cysteine sulfoxide, a reactive metabolite of S-(1,2,2-trichlorovinyl)-L-cysteine formed in rat liver and kidney microsomes, is a potent nephrotoxicant. J. Pharmacol. Exp. Ther. 321(3): 1095-1101.

EPA (U.S. Environmental Protection Agency). 1991. Guidelines for Developmental Toxicity Risk Assessment. EPA/600/FR-91/001. Risk Assessment Forum, U.S. Environmental Protection Agency, Washington, DC. December 1991 [online]. Available: http://www.epa.gov/raf/publications/pdfs/DEVTOX.PDF [accessed Oct. 13, 2009].

EPA (U.S. Environmental Protection Agency). 1992. Guidance on Risk Characterization for Risk Managers and Risk Assessors. Memorandum to Assistant Administrators, and Regional Administrators, from F. Henry Habicht, Deputy Administrator, Office of the Administrator, Washington, DC. February 26, 1992 [online]. Available: http://www.epa.gov/oswer/riskassessment/pdf/habicht.pdf [accessed Oct. 13, 2009].

EPA (U.S. Environmental Protection Agency). 1995. Guidance for Risk Characterization. Science Policy Council, U.S. Environmental Protection Agency, Washington, DC. February 1995 [online]. Available: http://www.epa.gov/osa/spc/pdfs/rcguide.pdf [accessed Oct. 13, 2008].

EPA (U.S. Environmental Protection Agency). 1996. Guidelines for Reproductive Toxicity Risk Assessment. EPA/630/R-96/009. Risk Assessment Forum, U.S. Environmental Protection Agency, Washington, DC. October 1996 [online]. Available: http://www.epa.gov/ncea/raf/pdfs/repro51.pdf [accessed Oct. 13, 2009].

EPA (U.S. Environmental Protection Agency). 1997. Guiding Principles for Monte Carlo Analysis. EPA/630/R-97/001. Risk Assessment Forum, U.S. Environmental Protection Agency, Washington, DC [online]. Available: http://www.epa.gov/NCEA/pdfs/montcarl.pdf [accessed Oct. 13, 2009].

EPA (U.S. Environmental Protection Agency). 1998. Guidance for Submission of Probabilistic Human Health Exposure Assessments to the Office of Pesticide Programs. Draft. Office of Pesticide Programs, U.S. Environmental Protection Agency. Nov. 4, 1998 [online]. Available: http://www.epa.gov/fedrgstr/EPA-PEST/1998/November/Day-05/6021.pdf [accessed Aug. 8, 2007].

EPA (U.S. Environmental Protection Agency). 2000. EPA Science Policy Council Handbook: Risk Characterization. EPA 100-B-00-002. Science Policy Council, U.S. Environmental Protection Agency, Washington, DC. December 2000 [online]. Available: http://www.epa.gov/osa/spc/pdfs/rchandbk.pdf [accessed Oct. 13, 2009].

EPA (U.S. Environmental Protection Agency). 2001. Risk Assessment Guidance for Superfund, Vol. III - Part A. Process for Conducting Probabilistic Risk Assessment. EPA 540-R-02-002. Office of Emergency and Remedial Response, U.S. Environmental Protection Agency, Washington, DC [online]. Available: http://www.epa.gov/oswer/riskassessment/rags3adt/ [accessed Aug. 8, 2009].

EPA (U.S. Environmental Protection Agency). 2002. A Review of the Reference Dose and Reference Concentration Processes. EPA/630/P-02/002F. Risk Assessment Forum, U.S. Environmental Protection Agency, Washington, DC [online]. Available: http://www.epa.gov/IRIS/RFD_FINAL%5B1%5D.pdf [accessed Oct. 22, 2009].

EPA (U.S. Environmental Protection Agency). 2005a. Guidelines for Carcinogen Risk Assessment. EPA/630/P-03/001B. Risk Assessment Forum, U.S. Environmental Protection Agency, Washington, DC. March 2005 [online]. Available: http://earth1.epa.gov/ttn/atw/cancer_guidelines_final_3-25-05.pdf [accessed Oct. 13, 2009].

EPA (U.S. Environmental Protection Agency). 2005b. Supplemental Guidance for Assessing Susceptibility from Early-Life Exposure to Carcinogens. EPA/630/R-03/003F. Risk Assessment Forum, U.S. Environmental Protection Agency, Washington, DC. March 2005 [online]. Available: http://earth1.epa.gov/ttn/atw/childrens_supplement_final.pdf [accessed Oct. 13, 2009].

EPA (U.S. Environmental Protection Agency). 2008. Toxicological Review of Tetrachloroethylene (Perchloroethylene) (CAS No. 127-18-4) in Support of Summary Information on the Integrated Risk Information System (IRIS). External Review Draft. U.S. Environmental Protection Agency, Washington, DC. June 2008 [online]. Available: http://cfpub.epa.gov/ncea/cfm/recordisplay.cfm?deid=192423 [accessed June 24, 2008].

EPA (U.S. Environmental Protection Agency). 2009. Using Probabilistic Methods to Enhance the Role of Risk Analysis in Decision-Making. External Review Draft. Prepared by Risk Assessment Forum PRA Technical Panel Working Groups, U.S. Environmental Protection Agency, Washington, DC.

EPA SAB (U.S. Environmental Protection Agency Science Advisory Board). 2007. Consultation on Enhancing Risk Assessment Practices and Updating EPA's Exposure Guidelines. EPA-SAB-07-003. Letter from Rebecca T. Parkin, Chair, Integrated Human Exposure, Committee and Environmental Health Committee, and Granger Morgan, Chair, Science Advisory Board, to Stephen L. Johnson, Administrator, U.S. Environmental Protection Agency, Washington, DC. February 28, 2007 [online]. Available: http://yosemite.epa.gov/sab/sabproduct.nsf/02ad90b136fc21ef85256eba00436459/55E1B2C78C6085EB8525729C00573A3E/$File/sab-07-003.pdf [accessed Oct. 13, 2009].

Epstein, D.L., G.A. Nolen, J.L. Randall, S.A. Christ, E.J. Read, J.A. Stober, and M.K. Smith. 1992. Cardiopathic effects of dichloroacetate in the fetal Long-Evans rat. Teratology 46(3):225-235.

Ericksson, P., D. von Rosen, H. Viberg, and A. Frederiksson. 2005. Developmental toxicology in the neonatal mouse: The use of randomly selected individuals as statistical unit compared to the litter in mice neonatally exposed to PBDE 99. Toxicologist 84(S-1):219-220.

Eskenazi, B., L. Fenster, M. Hudes, A.J. Wyrobek, D.F. Katz, J. Gerson, and D.M. Rempel. 1991a. A study of the effect of perchloroethylene exposure on the reproductive outcomes of wives of dry-cleaning workers. Am. J. Ind. Med. 20(5):593-600.

Eskenazi, B., A.J. Wyrobek, L. Fenster, D.F. Katz, M. Sadler, J. Lee, M. Hudes, and D.M. Rempel. 1991b. A study of the effect of perchloroethylene exposure on semen quality in dry cleaning workers. Am. J. Ind. Med. 20(5):575-591.

Evans, J.S., J.D. Graham, G.M. Gray, and R.L. Sielken. 1994. A distributional approach to characterizing low-dose cancer risk. Risk Anal. 14(1):25-34.

Ferroni, C., L. Selis, A. Mutti, D. Folli, E. Bergamaschi, and I. Franchini. 1992. Neurobehavioral and neuroendocrine effects of occupational exposure to perchloroethylene. Neurotoxicology 13(1):243-247.

Frantz, S.W., and P.G. Watanabe. 1983. Tetrachloroethylene: Balance and tissue distribution in male Sprague-Dawley rats by drinking-water administration. Toxicol. Appl. Pharmacol. 69(1):66-72.

Fredriksson, M., B.R. Danielsson, and P. Eriksson. 1993. Altered behaviour in adult mice orally exposed to tri- and tetrachloroethylene as neonates. Toxicol. Lett. 66(1):13-19.

Galloway, S.M., M.J. Armstrong, C. Reuben, S. Colman, B. Brown, C. Cannon, A.D. Bloom, F. Nakamura, M. Ahmed, S. Duk, J. Rimpo, B.H. Margolin, M.A. Resnick, B. Anderson, and E. Zeiger. 1987. Chromosome aberrations and sister chromatid exchanges in Chinese hamster ovary cells: Evaluations of 108 chemicals. Environ. Mol. Mutagen. 19(Suppl. 10):1-175.

Gearhart, J.M., D.A. Mahle, R.J. Greene, C.S. Seckel, C.D. Flemming, J.W. Fisher, and H.J. Clewell, III. 1993. Variability of physiologically based pharmacokinetic (PBPK) model parameters and their effects on PBPK model predictions in a risk assessment for perchloroethylene (PCE). Toxicol. Lett. 68(1-2):131-144.

Gennari, P., M. Naldi, R. Motta, M.C. Nucci, C. Giacomini, F.S. Violante, and G.B. Raffi. 1992. Gamma-glutamyltransferase isoenzyme pattern in workers exposed to tetrachloroethylene. Am. J. Ind. Med. 21(5):661-671.

Gobba, F. 2003. Occupational exposure to chemicals and sensory organs: A neglected research field. Neurotoxicology 24(4-5):675-691.

Gobba, F., E. Righi, G. Fantuzzi, G. Predieri, L. Cavazzuti, and G. Aggazzotti. 1998. Two-year evolution of perchloroethylene-induced color-vision loss. Arch. Environ. Health 53(3):196-198.

Goldsworthy, T.L., and J.A. Popp. 1987. Chlorinated hydrocarbon-induced peroxisomal enzyme activity in relation to species and organ carcinogenicity. Toxicol. Appl. Pharmacol. 88(2):225-233.

Goldsworthy, T.L., T. Smith-Oliver, D.J. Loury, J.A. Popp, and B.E. Butterworth. 1988. Assessment of chlorinated hydrocarbon-induced genotoxicity and cell replication in rat kidney cells. Environ. Mol. Mutagen. 11(Suppl. 11):39 [Abstract 92].

Grandjean, P., and P.J. Landrigan. 2006. Developmental neurotoxicity of industrial chemicals. Lancet 368(9553):2167-2178.

Green, T., J. Odum, J.A. Nash, and J.R. Foster. 1990. Perchloroethylene-induced rat kidney tumors: An investigation of the mechanisms involved and their relevance to humans. Toxicol. Appl. Pharmacol. 103(1):77-89.

Guo, C., and K.E. McMartin. 2005. The cytotoxicity of oxalate, a metabolite of ethylene glycol, is due to calcium oxalate monohydrate formation. Toxicology 208(3):347-355.

Hake, C.L., and R.D. Stewart. 1977. Human exposure to tetrachloroethylene: Inhalation and skin contact. Environ. Health Perspect. 21:231-238.

Hardin, B.D., G.P. Bond, M.R. Sikov, F.D. Andrew, R.P. Beliles, and R.W. Niemeier. 1981. Testing of selected workplace chemicals for teratogenic potential. Scand. J. Work Environ. Health 7(Suppl. 4):66-75.

Hartmann, A., and G. Speit. 1995. Genotoxic effects of chemicals in the single cell gel (SCG) test with human blood cells in relation to the induction of sister-chromatid exchanges (SCE). Mutat. Res. 346(1):49-56.

Haseman, J.K., J.R. Hailey, and R.W. Morris. 1998. Spontaneous neoplasm incidences in Fischer 344 rats and B6C3F1 mice in two-year carcinogenicity studies: a National Toxicology Program update. Toxicol. Pathol. 26(3):428-441.

Hill, B. 1965. The environment and disease: Association or causation? Proc. Royal Soc. Med. 58(May):295-300.

Honma, T., A. Sudo, M. Miyagawa, M. Sato, and H. Hasegawa. 1980. Effects of exposure to trichloroethylene and tetrachloroethylene on the contents of acetylcholine, dopamine, norepinephrine and serotonin in rat brain. Ind. Health 18(4):171-178.

Houston, J.B., and D.J. Carlile. 1997. Prediction of hepatic clearance from microsomes, hepatocytes, and liver slices. Drug. Metab. Rev. 29(4):891-922.

IARC (International Agency for Research on Cancer). 2004. Some Drinking-Water Disinfectants and Contaminants, Including Arsenic. IARC Monographs on the Evaluation of Carcinogenic Risks to Humans, Vol. 84. Lyon: IARC.

IEc (Industrial Economics Inc.). 2006. Expanded Expert Judgment Assessment of the Concentration-Response Relationship Between PM2.5 Exposure and Mortality, Final Report. Prepared by Industrial Economics, Incorporated, Cambridge, MA, for Office of Air Quality Planning and Standards, U.S. Environmental Protection Agency, Research Triangle Park, NC. September 2006 [online]. Available: http://www.epa.gov/ttn/ecas/regdata/Uncertainty/pm_ee_report.pdf [accessed Oct. 14, 2009].

Ikeda, M., A. Koizumi, T. Watanabe, A. Endo, and K. Sato. 1980a. Cytogenetic and cytokinetic investigations on lymphocyes from workers occupationally exposed to tetrachloroethylene. Toxicol. Lett. 5(3-4):251-256.

Ikeda, M., A. Koizumi, T. Watanabe, A. Endo, and K. Sato. 1980b. Cytogenetic and cytokinetic investigations on lymphocytes from workers occupationally exposed to tetrachloroethylene. Dev. Toxicol. Environ. Sci. 8:347-350.

Ishmael, J., and P.H. Dugard. 2006. A review of perchloroethylene and rat mononuclear cell leukemia. Regul. Toxicol. Pharmacol. 45(2):178-184.

Ji, J., and K. Hemminki. 2005. Occurrences of leukemia subtypes by socioeconomic and occupational group. J. Occup. Environ. Med. 47(11):1131-1140.

Ji, J., and K. Hemminki. 2006. Socioeconomic/occupational risk factors for lymphoproliferative diseases in Sweden. Ann. Epidemiol. 16(5):370-376.

JISA (Japanese Industrial Safety Association). 1993. Carcinogenicity Study of Tetrachloroethylene by Inhalation in Rats and Mice. Data No. 3-1. Kanagawa, Japan: Japanese Industrial Safety Association.

Johnson, P.D., B.V. Dawson, and S.J. Goldberg. 1998. Cardiac teratogenicity of trichloroethylene metabolites. J. Am. Coll. Cardiol. 32(2):540-545.

Karlsson, J.E., L.E. Rosengren, P. Kjellstrand, and K.G. Haglid. 1987. Effects of low-dose inhalation of three chlorinated aliphatic organic solvents on deoxyribonucleic acid in gerbil brain. Scan. J. Work Environ. Health 13(5):453-458.`

Ketcha, M.M., D.K. Stevens, D.A. Warren, C.T. Bishop, and W.T. Brashear. 1996. Conversion of trichloroacetic acid to dichloroacetic acid in biological samples. J. Anal. Toxicol. 20(4):236-241.

King-Herbert, A., and K. Thayer. 2006. NTP Workshop: Animal models for the NTP rodent bioassay: Stocks and strains- should we switch? Toxicol. Pathol. 34(6):802-805.

Kjellstrand, P., B. Holmquist, M. Kanje, P. Alm, S. Romare, I. Jonsson, L. Mansson, and M. Bjerkemo. 1984. Perchloroethylene: Effects on body and organ weights and plasma butyrylcholinesterase activity in mice. Acta Pharmacol. Toxicol. 54(5): 414-424.

Kjellstrand, P., B. Holmquist, I. Jonsson, S. Romare, and L. Mansson. 1985. Effects of organic solvents on motor activity in mice. Toxicology 35(1):35-46.

Klaassen, C.D., and G.L. Plaa. 1966. Relative effects of various chlorinated hydrocarbons on liver and kidney function in mice. Toxicol. Appl. Pharmacol. 9(1):139-151.

Klaassen, C.D., and G.L. Plaa. 1967. Relative effects of various chlorinated hydrocarbons on liver and kidney function in dogs. Toxicol. Appl. Pharmacol. 10(1):119-131.

Krupnick, A., R. Morgenstern, M. Batz, P. Nelson, D. Burtraw, J. Shih, and M. McWilliams. 2006. Not a Sure Thing: Making Regulatory Choices Under Uncertainty. Washington, DC: Resources for the Future [online]. Available: http://www.rff.org/RFF/Documents/RFF-Rpt-RegulatoryChoices.pdf [accessed Oct. 14, 2009].

Kyrklund, T., and K. Haglid. 1991. Brain lipid composition in guinea pigs after intrauterine exposure to perchloroethylene. Pharmacol. Toxicol. 68(2):146-148.

Kyrklund, T., C. Alling, P. Kjellstrand, and K.G. Haglid. 1984. Chronic effects of perchloroethylene on the composition of lipid and acyl groups in cerebral cortex and hippocampus of the gerbil. Toxicol. Lett. 22(3):343-349.

Kyrklund, T., P. Kjellstrand, and K.G. Haglid. 1987. Lipid concentration and fatty acid pattern of the gerbil brain after exposure to perchloroethylene. Arch. Toxicol. 60(5):397-400.

Kyyronen, P., H. Taskinen, M.L. Lindbohm, K. Hemminki, and O.P. Heinonen. 1989. Spontaneous abortions and congenital malformations among women exposed to tetrachloroethylene in dry cleaning. J. Epidemiol. Community Health 43(4):346-351.

Larson, J.L., and R.J. Bull. 1992. Metabolism and lipoperoxidative activity of trichloroacetate and dichloroacetate in rats and mice. Toxicol. Appl. Pharmacol. 115(2):268-277.

Lash, L.H., and J.C. Parker. 2001. Hepatic and renal toxicities associated with perchloroethylene. Pharmacol. Rev. 53(2):177-208.

Lash, L.H., W. Qian, D.A. Putt, K. Desai, A.A. Elfarra, A.R. Sicuri, and J.C. Parker. 1998. Glutathione conjugation of perchloroethylene in rats and mice in vitro: Sex-, species-, and tissue-dependent differences. Toxicol. Appl. Pharmacol. 150(1):49-57.

Lash, L.H., D.A. Putt, P. Huang, S.E. Hueni, and J.C. Parker. 2007. Modulation of hepatic and renal metabolism and toxicity of trichloroethylene and perchloroethylene by alterations in status of cytochrome P450 and glutathione. Toxicology 235(1-2):11-26.

Lauwerys, R., J. Herbrand, J.P. Buchet, A. Bernard, and J. Gaussin. 1983. Health surveillance of workers exposed to tetrachloroethylene in dry-cleaning shops. Int. Arch. Occup. Environ. Health 52(1):69-77.

Lezak, M.D., D.B. Howieson, D.W. Loring, H.J. Hannay, and J.S. Fischer, eds. 2004. Neuropsychological Assessment, 4th Ed. New York: Oxford University Press.

Lock, E.A., Y. Sani, R.B. Moore, M.B. Finkelstein, M.W. Anders, and A.A. Seawright. 1996. Bone marrow and renal injury associated with haloalkene cysteine conjugates in calves. Arch. Toxicol. 70(10):607-619.

Lukaszewski, T. 1979. Acute tetrachloroethylene fatality. Clin. Toxicol. 15(4):411-415.

Lynge, E., and L. Thygesen. 1990. Primary liver cancer among women in laundry and dry-cleaning work in Denmark. Scand. J. Work Environ. Health 16(2):108-112.

Lynge, E., B. Carstensen, and O. Andersen. 1995. Primary liver cancer and renal cell carcinoma in laundry and dry-cleaning workers in Denmark. Scand. J. Work Environ. Health 21(4):293-295.

Lynge, E., A. Andersen, L. Rylander, H. Tinnerberg, M.L. Lindbohm, E. Pukkala, P. Romundstad, P. Jensen, L.B. Clausen, and K. Johansen. 2006. Cancer in persons working in dry cleaning in the Nordic countries. Environ. Health Perspect. 114(2):213-219.

Maltoni, C., and G. Cotti. 1986. Results of long-term carcinogenicity bioassays of tetrachloroethylene on Sprague-Dawley rats administered by ingestion. Acta Oncol. 7(1):11-26.

Mandel, J.S., J.K. McLaughlin, B. Schlehofer, A. Mellemgaard, U. Helmert, P. Lindblad, M. McCredie, and H.O. Adami. 1995. International renal-cell cancer study. IV. Occupation. Int. J. Cancer 61(5):601-605.

Marth, E. 1987. Metabolic changes following oral exposure to tetrachloroethylene in subtoxic concentrations. Arch. Toxicol. 60(4):293-299.

Marth, E., D. Stunzner, H. Binder, and J.R. Mose. 1985. Tetrachloroethylene: Effect of low concentrations of 1,1,2,2-tetrachloroethylene (tetrachloroethylenehloroethylene) on organisms in the mouse. II. Study of tetrachloroethylene in various organs and demonstration of histological changes in examined organs [in German]. Zentralbl. Bakteriol. Mikrobiol. Hyg. B. 181(6):541-547.

Marth, E., D. Stunzner, M. Kock, and J.R. Mose. 1989. Toxicokinetics of chlorinated hydrocarbons. J. Hyg. Epidemiol. Microbiol. Immunol. 33(4 suppl.):514-520.

Mattsson, J.L., R.R. Albee, B.L. Yano, G. Bradley, and P.J. Spencer. 1998. Neurotoxicologic examination of rats exposed to 1,1,2, 2-tetrachloroethylene (perchloroethylene) vapor for 13 weeks. Neurotoxicol. Teratol. 20(1):83-98.

Mazullo, M, S. Grilli, G. Lattanzi, G. Prodi, M.P. Turina, and A. Colacci. 1987. Evidence of DNA binding activity of perchloroethylene. Res. Commun. Chem. Pathol. Pharmacol. 58(2):215-235.

McGoldrick, T.A., E.A. Lock, V. Rodilla, and G.M. Hawksworth. 2003. Renal cysteine conjugate C-S lyase mediated toxicity of halogenated alkenes in primary cultures of human and rat proximal tubular cells. Arch. Toxicol. 77(7):365-370.

Meek, M.E. 2008. Recent developments in frameworks to consider human relevance of hypothesized modes of action for tumours in animals. Environ. Mol. Mutagen. 49(2):110-116.

Mellemgaard, A., G. Engholm, J.K. McLaughlin, and J.H. Olsen. 1994. Occupational risk factors for renal-cell carcinoma in Denmark. Scand. J. Work Environ. Health 20(3):160-165.

Mihic, S.J. 1999. Acute effects of ethanol on GABAA and glycine receptor function. Neurochem. Int. 35(2):115-123.

Miligi, L., A.S. Costantini, A. Benvenuti, D. Kriebel, V. Bolejack, R. Tumino, V. Ramazzotti, S. Rodella, E. Stagnaro, P. Crosignani, D. Amadori, D. Mirabelli, L. Sommani, I. Belletti, L. Troschel, L. Romeo, G. Miceli, G.A. Tozzie, I. Mendico, and P. Vineis. 2006. Occupational exposure to solvents and the risk of lymphomas. Epidemiology 17(5):552-561.

Miller, R.E., and F.P. Guengerich. 1982. Oxidation of trichloroethylene by liver microsomal cytochrome P-450: Evidence for chlorine migration in a transition state not involving trichloroethylene oxide. Biochemistry 21(5):1090-1097.

Milman, H.A., D.L. Story, E.S. Riccio, A. Sivak, A.S. Tu, G.M. Williams, C. Tong, and C.A. Tyson. 1988. Rat liver foci and in vitro assays to detect initiating and promoting effects of chlorinated ethanes and ethylenes. Ann. N.Y. Acad. Sci. 534:521-530.

Monster, A.C., G. Boersma, and H. Steenweg. 1979. Kinetics of tetrachloroethylene in volunteers: influence of exposure concentration and work load. Int. Arch. Occup. Environ. Med. 42(3-4):303-309.

Moore, M.M., and K. Harrington-Brock. 2000. Mutagenicity of trichloroethylene and its metabolites: Implications for the risk assessment of trichloroethylene. Environ. Health Perspect. 108(Suppl. 2):215-223.

Morgan, M.G., and M. Henrion. 1990. Uncertainty: A Guide to Dealing with Uncertainty in Quantitative Risk and Policy Analysis. New York: Cambridge University Press.

Morton, W., and D. Marjanovic. 1984. Leukemia incidence by occupation in the Portland-Vancouver metropolitan area. Am. J. Ind. Med. 6(3):185-205.

Moser, V.C., B.M. Cheek, and R.C. MacPhail. 1995. A multidisciplinary approach to toxicological screening: III. Neurobehavioral toxicity. J. Toxicol. Environ. Health 45(2):173-210.

Motohashi, Y., Y. Miyazaki, and T. Takano. 1993. Assessment of behavioral effects of tetrachloroethylene using a set of time-series analyses. Neurtoxicol. Teratol. 15(1):3-10.

Muller, M., G. Birner, and W. Dekant. 1998a. Reactivity of haloketenes and halothioketenes with nucleobases: Chemical characterization of reaction products. Chem. Res. Toxicol. 11(5):454-463.

Muller, M., G. Birner, M. Sander, and W. Dekant. 1998b. Reactivity of haloketenes and halothioketenes with nucleobases: Reactions in vitro with DNA. Chem. Res. Toxicol. 11(5):464-470.

Murakami, K., and K. Horikawa. 1995. The induction of micronuclei in mice hepatocytes and reticulocytes by tetrachloroethylene. Chemosphere 31(7):3733-3739.

Nagano, K., T. Nishizawa, S. Yamamoto, and T. Matsushima. 1998. Inhalation carcinogenesis studies of six halogenated hydrocarbons in rats and mice. Pp. 741-746 in Advances in the Prevention of Occupational Respiratory Diseases: Proceedings of the 9th International Conference on Occupational Respiratory Disease, Kyoto, Japan, 13-16 October 1997, K. Chiyotani, Y. Hosoda, and Y. Aizawa, eds. Amsterdam: Elsevier.

Nakajima, T., Y. Kamijo, N. Usuda, Y. Liang, Y. Fukushima, K. Kametani, F.J. Gonzalez, and T. Aoyama. 2000. Sex-dependent regulation of hepatic peroxisome proliferation in mice by trichloroethylene via peroxisome proliferator-activated receptor alpha (PPARalpha). Carcinogenesis 21(4):677-682.

NCI (National Cancer Institute). 1977. Bioassay of Tetrachloroethylene for Possible Carcinogenicity. NCI-CG-TR-13. DHEW(NIH)77-813. U.S. Department of Health, Education, and Welfare, Public Health Service, National Institute of Health, Bethesda, MD [online]. Available: http://ntp.niehs.nih.gov/ntp/htdocs/LT_rpts/tr013.pdf [accessed Oct. 14, 2009].

Nelson, B.K., B.J. Taylor, J.V. Setzer, and J.W. Hornung. 1979. Behavioral teratology of perchloroethylene in rats. J. Environ. Pathol. Toxicol. 3(1-2):233-250.

NRC (National Research Council). 1983. Risk Assessment in the Federal Government: Managing the Process. Washington, DC: National Academy Press.

NRC (National Research Council). 1989. Improving Risk Communication. Washington, DC: National Academy Press.

NRC (National Research Council). 1993. Issues in Risk Assessment. Washington, DC: National Academy Press.

NRC (National Research Council). 1994. Science and Judgment in Risk Assessment. Washington, DC: National Academy Press.

NRC (National Research Council). 2002. Estimating the Public Health Benefits of Proposed Air Pollution Regulations. Washington, DC: The National Academies Press.

NRC (National Research Council). 2006a. Assessing the Human Health Risks of Trichloroethylene. Washington, DC: The National Academies Press.

NRC (National Research Council). 2006b. Health Risks from Dioxin and Related Compounds: Evaluation of the EPA Reassessment. Washington, DC: The National Academies Press.

NRC (National Research Council). 2007. Models in Environmental Regulatory Decision Making. Washington, DC: The National Academies Press.

NRC (National Research Council). 2009. Science and Decisions: Advancing Risk Assessment. Washington, DC: The National Academies Press.

NTP (National Toxicology Program). 1986. Toxicology and Carcinogenesis Studies of Tetrachloroethylene (Perchloroethylene) CAS No. 127-18-4) in F344/N Rats and $B6C3F_1$ Mice (Inhalation Studies). NTP TR 311. NIH Publication No. 86-2567. U.S. Department of Health and Human Services, Public Health Service, National Institutes of Health, Research Triangle Park, NC.

Odum, J., T., Green, J.R. Foster, and P.M. Hext. 1988. The role of trichloroacetic acid and peroxisome proliferation in the differences in carcinogenicity of perchloroethylene in the mouse and rat. Toxicol. Appl. Pharmacol. 92(1):103-112.

Olsen, J., K. Hemminki, G. Ahlborg, T. Bjerkedal, P. Kyyronen, H. Taskinen, M.L. Lindbohm, O.P. Heinonen, L. Brandt, H. Kolstad, B.A. Halvorsen, and J. Egenaes. 1990. Low birth weight, congenital malformations, and spontaneous abortions among dry-cleaning workers in Scandinavia. Scand. J. Work Environ. Health 16(3):163-168.

Oshiro, W.M., Q.T. Krantz, and P.J. Bushnell. 2008. Characterization of the effects of inhaled perchloroethylene on sustained attention in rats performing a visual signal detection task. Neurotoxicol. Teratol. 30(3):167-174.

Ozkaynak, H., H.C. Frey, J. Burke, and R.W. Pinder. 2008. Analysis of coupled model uncertainties in source to dose modeling of human exposures to ambient air pollution: A PM2.5 case-study. Atmos. Environ. 43(9):1641-1649.

Pahler, A., G. Birner, J. Parker, and W. Dekant. 1998. Generation of antibodies to di- and trichloroacetylated proteins and immunochemical detection of protein adducts in rats treated with perchloroethene. Chem. Res. Toxicol. 11(9):995-1004.

Pahler, A., J. Parker, and W. Dekant. 1999. Dose-dependent protein adduct formation in kidney, liver, and blood of rats and in human blood after perchloroethene inhalation. Toxicol. Sci. 48(1):5-13.

Pang, K.S. 2009. Safety testing of metabolites: Expectations and outcomes. Chem. Biol. Interact. 179(1):45-49.

Parrish, J.M., E.W. Austin, D.K. Stevens, D.H. Kinder, and R.J. Bull. 1996. Haloacetate-induced oxidative damage to DNA in the liver of male B6C3F1 mice. Toxicology 100(1-3):103-111.

Partanen, T., P. Heikkila, S. Hernberg, T. Kauppinen, G. Moneta, and A. Ojajarvi. 1991. Renal cell cancer and occupational exposure to chemical agents. Scand. J. Work Environ. Health 17(4):231-239.

PCCRARM (Presidential/Congressional Commission on Risk Assessment and Risk Management). 1997. Framework for Environmental Health Risk Management Final Report, Volume 1 [online]. Available: http://www.riskworld.com/nreports/1997/risk-rpt/pdf/EPAJAN.PDF [accessed Oct. 14, 2009].

Pegg, D.G., J.A. Zempel, W.H. Braun, and P.G. Watanabe. 1979. Disposition of tetrachloro(14C)ethylene following oral and inhalation exposure in rats. Toxicol. Appl. Pharmacol. 51(3):456-474.

Perocco, P., S. Bolognesi, and W. Alberghini. 1983. Toxic activity of seventeen industrial solvents and halogentated compounds on human lymphocytes cultured in vitro. Toxicol. Lett. 16(1-2):69-75.

Pesch, B., J. Haerting, U. Ranft, A. Klimpel, B. Oelschlagel, W. Schill, and MURC Study Group. 2000. Occupational risk factors for renal cell carcinoma: Agent-specific results from a case-control study in Germany. Int. J. Epidemiol. 29(6):1014-1024.

Philip, B.K., M.M. Mumtaz, J.R. Latendresse, and H.M. Mehendale. 2007. Impact of repeated exposure on toxicity of perchloroethylene in Swiss Webster mice. Toxicology 232(1-2):1-14.

Potter, C.L., L.W. Chang, A.B. DeAngelo, and F.B. Daniel. 1996. Effects of four trihalomethanes on DNA strand breaks, renal hyaline droplet formation and serum testosterone in male F-344 rats. Cancer Lett. 106(2):235-242.

Rao, V.R., and D.R. Brown. 1993. A physiologically based pharmacokinetic assessment of tetrachloroethylene in groundwater for a bathing and showering determination. Risk Anal. 13(1):37-49.

Reitz, R.H., M.L. Gargas, A.L. Mendrala, and A.M. Schumann. 1996. In vivo and in vitro studies of perchloroethylene metabolism for physiologically based pharmacokinetic modeling in rats, mice, and humans. Toxicol. Appl. Pharmacol. 136(2):289-306.

Ripp, S.L., L.H. Overby, R.M. Philpot, and A.A. Elfarra. 1997. Oxidation of cysteine *S*-conjugates by rabbit liver microsomes and cDNA-expressed flavin-containing monooxygenases: Studies with *S*-(1,2-dichlorovinyl)-L-cysteine, *S*-(1,2,2-trichlorovinyl)-L-cysteine, *S*-allyl-L-cysteine, and *S*-benzyl-L-cysteine. Mol. Pharmacol. 51(3):507-515.

Rosengren, L.E., P. Kjellstrand, and K.G. Haglid. 1986. Tetrachloroethylene: Levels of DNA and S-100 in the gerbil CNS after chronic exposure. Neurobehav. Toxicol. Teratol. 8(2):201-206.

Ruder, A.M., E.M. Ward, and D.P. Brown. 2001. Mortality in dry-cleaning workers: An update. Am. J. Ind. Med. 39(2):121-132.

Savolainen, H., P. Pfaffli, M. Tengen, and H. Vainio. 1977. Biochemical and behavioral effects of inhalation exposure to tetrachloroethylene and dichloromethane. J. Neuropathol. Exp. Neurol. 36(6):941-949.
Schlehofer, B., C. Heuer, M. Blettner, D. Niehoff, and J. Wahrendorf. 1995. Occupation, smoking and demographic factors, and renal cell carcinoma in Germany. Int. J. Epidemiol. 24(1):51-57.
Schreiber, J.S., H.K. Hudnell, A.M. Geller, D.E. House, K.M. Aldous, M.S. Force, K. Langguth, E.J. Prohonic, and J.C. Parker. 2002. Apartment residents' and day care workers' exposures to tetrachloroethylene and deficits in visual contrast sensitivity. Environ. Health Perspect. 110(7):655-664.
Schumann, A.M., J.F. Quast, and P.G. Watanabe. 1980. The pharmacokinetics and macromolecular interactions of perchloroethylene in mice and rats as related to oncogenicity. Toxicol. Appl. Pharmacol. 55(2):207-219.
Schwetz, B.A., K.J. Leong, and P.J. Gehring. 1975. The effect of maternally inhaled trichloroethylene, perchloroethylene, methyl chloroform, and methylene chloride on embryonal and fetal development in mice and rats. Toxicol. Appl. Pharmacol. 32(1):84-96.
Scott, W.J., Jr, M.D. Collins, and H. Nau. 1994. Pharmacokinetic determinants of embryotoxicity in rats associated with organic acids. Environ. Health Perspect. 102 (Suppl. 11):97-101.
Seeber, A. 1989. Neurobehavioral toxicity of long-term exposure to tetrachloroethylene. Neurotoxicol. Teratol. 11(6):570-583.
Seidel, H.J., L. Weber, and E. Barthel. 1992. Hematological toxicity of tetrachloroethylene in mice. Arch. Toxicol. 66(3):228-230.
Shafer, T.J., P.J. Bushnell, V.A. Benignus, and J.J. Woodward. 2005. Perturbation of voltage-sensitive Ca2+ channel function by volatile organic solvents. J. Pharmacol. Exp. Ther. 315(3):1109-1118.
Shimada, T. A.F. Swanson, P. Leber, and G.M. Williams. 1985. Activities of chlorinated ethane and ethylene compounds in the Salmonella/rat microsome mutagenesis and rat hepatocyte/DNA repair assays under vapor phase exposure conditions. Cell Biol. Toxicol. 1(3):159-179.
Slikker, W., Jr. 1994. Principles of developmental neurotoxicology. Neurotoxicology 15 (1):11-16.
Small, M.J. 2008. Methods for assessing uncertainty in fundamental assumptions and associated models for cancer risk assessment. Risk Anal. 28(5):1289-1307.
Smith, M.K., J.L. Randall, E.J. Read, and J.A. Stober. 1989. Teratogenic activity of trichloroacetic acid in the rat. Teratology 40(5):445-451.
Smith, M.K., J.L. Randall, E.J. Read, and J.A. Stober. 1992. Developmental toxicity of dichloroacetate in the rat. Teratology 46(3):217-223.
Sofuni, T., M. Hayashi, A. Matsuoka, M. Sawada, M. Hatanaka, and M. Ishidate, Jr. 1985. Mutagenicity tests on organic chemical contaminants in city water and related compounds. II. Chromosome aberration tests in cultured mammanlian cells [in Japanese]. Eisei Shikenjo Hokoku 103:64-75.
Sonnenfeld, N., I. Hertz-Picciotto, and W.E. Kaye. 2001. Tetrachloroethylene in drinking water and birth outcomes at the U.S. Marine Corps Base at Camp Lejeune, North Carolina. Am. J. Epidemiol. 154(10):902-908.
Spinatonda, G., R. Colombo, E.M. Capodaglio, M. Imbriani, C. Pasetti, G. Minuco, and P. Pinelli. 1997. Process of speech production: Application in a group of subjects chronically exposed to organic solvents (II) [in Italian]. G. Ital. Med. Lav. Ergon. 19(3):85-88.

Spirtas, R., P.A. Stewart, J.S. Lee, D.E. Marano, C.D. Forbes, D.J. Grauman, H.M. Pettigrew, A. Blair, R.N. Hoover, and J.L. Cohen. 1991. Retrospective cohort mortality study of workers at an aircraft maintenance facility. I. Epidemiological results. Br. J. Ind. Med. 48(8):515-530.

Stewart, R.D., and H.C. Dodd. 1964. Absorption of carbon tetrachloride, trichloroethylene, tetrachloroethylene, methylene chloride, and 1,1,1-trichloroethane through the human skin. Am. Ind. Hyg. Assoc. J. 25:439-446.

Storer, R.D., T.W. McKelvey, A.R. Kraynak, M.C. Elia, J.E. Barnum, L.S. Harmon, W.W. Nichols, and J.G. DeLuca. 1996. Revalidation of the in vitro alkaline elution/rat hepatocyte assay for DNA damage: Improved criteria for assessment of cytotoxicity and genotoxicity and results for 81 compounds. Mutat. Res. 368(2):59-101.

Storm, J.E., and K.A. Mazor. 2004. Update of residential tetrachloroethylene exposure and decreases in visual contrast sensitivity. Environ. Health Perspect. 112(15):A8 62-A864.

Stromberg, P.C. 1985. Large granular lymphocyte leukemia in F344 rats: A model for human T gamma lymphoma, malignant histiocytosis, and T-cell chronic lymphocytic leukemia. Am. J. Pathol. 119(3):517-519.

Sweeney, L.M., C.R. Kirman, M.L. Gargas, and P.H. Dugard. 2009. Contribution of trichloroacetic acid to liver tumors observed in perchloroethylene (perc)-exposed mice. Toxicology 260(1-3):77-83.

Szakmary, E., G. Ungvary, and E. Tatrai. 1997. The offspring-damaging effect of tetrachloroethylene in rats, mice, and rabbits. Cent. Eur. J. Occup. Environ. Med. 3(1): 31-39.

Thomas, J., J.K. Haseman, J.I. Goodman, J.M. Ward, T.P. Loughran, Jr., and P.J. Spencer. 2007. A review of large granular lymphocytic leukemia in Fischer 344 rats as an initial step toward evaluating the implication of the endpoint to human cancer risk assessment. Toxicol. Sci. 99(1): 3-19.

Tinston, D.J. 1994. Perchloroethylene: A Multigeneration Inhalation Study in the Rat. Report No. CTL/P/4097. Zeneca Central Toxicology Laboratory, Alderley Park, Macclesfield, Cheshire, UK. November 3, 1994.

Toraason, M., J. Clark, D. Dankovic, P. Mathias, S. Skaggs, C. Walker, and D. Werren. 1999. Oxidative stress and DNA damage in Fischer rats following acute exposure to trichloroethylene or perchloroethylene. Toxicology 138(1):43-53.

Toraason, M., M.A. Butler, A. Ruder, C. Forrester, L. Taylor, D.L. Ashley, P. Mathias, K.L. Marlow, K.L. Cheever, E. Krieg, and H. Wey. 2003. Effect of perchloroethylene, smoking, and race on oxidative DNA damage in female dry cleaners. Mutat. Res. 539(1-2):9-18.

Travier, N., G. Gridley, A.J. De Roos, N. Plato, T. Moradi, and P. Boffetta. 2002. Cancer incidence of dry cleaning, laundry, and ironing workers in Sweden. Scand. J. Work Environ. Health 28(5):341-348.

Tu, A.S., T.A. Murray, K.M. Hatch, A. Sivak, and H.A. Milman. 1985. In vitro transformation of BALB/c-3T3 cells by chlorinated ethanes and ethylenes. Cancer Lett. 28(1):85-92.

Umezu, T., J. Yonemoto, Y. Soma, and T. Miura. 1997. Behavioral effects of trichloroethylene and tetrachloroethylene in mice. Pharmacol. Biochem. Behav. 58(3):665-671.

Vamvakas, S., W. Dekant, and D. Henschler. 1989a. Genotoxocity of haloalkene and haloalkane glutathione S-conjugates in porcine kidney cells. Toxicol. In Vitro 3(2): 151-156.

Vamvakas, S., A. Kochling, H. Berthold, and W. Dekant. 1989b. Cytotoxicity of cysteine S-conjugates: Structure-activity relationships. Chem. Biol. Interact. 71(1):79-90.

Vamvakas, S., W. Dekant, and D. Henschler. 1989c. Assessment of unscheduled DNA synthesis in a cultured line of renal epithelial cells exposed to cysteine S-conjugates of haloalkenes and haloalkanes. Mutat. Res. 222(4):329-335.

Vamvakas, S., M. Herkenhoff, W. Dekant, and D. Henschler. 1989d. Mutagenicity of tetrachloroethylene in the Ames test—metabolic activation by conjugation with glutathione. J. Biochem. Toxicol. 4(1):21-27.

Vaughan, T.L., P.A. Stewart, S. Davis, and D.B. Thomas. 1997. Work in dry-cleaning and the incidence of cancer of the oral cavity, larynx, and esophagus. Occup. Environ. Med. 54(9):692-695.

Verplanke, A.J., M.H. Leummens, and R.F. Herber. 1999. Occupational exposure to tetrachloroethylene and its effects on the kidneys. J. Occup. Environ. Med. 41(1):11-16.

Volkel, W., and W. Dekant. 1998. Chlorothioketene, the ultimate reactive intermediate formed by cysteine conjugate beta-lyase-mediated cleavage of the trichloroethene metabolite *S*-(1,2-dichlorovinyl)-L-cysteine, forms cytosine adducts in organic solvents, but not in aqueous solution. Chem. Res. Toxicol. 11(9):1082-1088.

Volkel, W., M. Friedewald, E. Lederer, A. Pahler, J. Parker, and W. Dekant. 1998. Biotransformation of perchloroethene: Dose-dependent excretion of trichloroacetic acid, dichloroacetic acid, and N-acetyl-S-(trichlorovinyl)-cysteine in rats and humans after inhalation. Toxicol. Appl. Pharmacol. 153(1):20-27.

Von Tungeln, L.S., P. Yi, T.J. Bucci, V.M. Samokyszyn, M.W. Chou, F.F. Kadlubr, and P.P. Fu. 2002. Tumorigenicity of chloral hydrate, trichloroacetic acid, trichloroethanol, malondialdehyde, 4-hydroxy-2-nonenal, crotonaldehyde, and acrolein in the B6C3F(1) neonatal mouse. Cancer Lett. 185(1):13-19.

Vose, D. 2008. Risk Analysis: A Quantitative Guide, 3rd Ed. Chichester, England: John Wiley and Sons.

Vyskocil, A., S. Emminger, J. Tejral, Z. Fiala, E. Ettlerova, and A. Cermanova. 1990. Study on kidney function in female workers exposed to perchloroethylene. Hum. Exp. Toxicol. 9(6):377-380.

Walles, S.A. 1986. Inducation of single-strand breaks in DNA of mice by trichloroethylene and tetrachloroethylene. Toxicol. Lett. 31(1):31-35.

Wang, J.L., W.L. Chen, S.Y. Tsai, P.Y. Sung, and R.N. Huang. 2001. An in vitro model for evaluation of vaporous toxicity of trichloroethylene and tetrachloroethylene to CHO-K1 cells. Chem. Biol. Interact. 137(2):139-154.

Wang, S., J.E. Karlsson, T. Kyrklund, and K. Haglid. 1993. Perchloroethylene-induced reduction in glial and neuronal cell marker proteins in rat brain. Pharmacol. Toxicol. 72(4-5):273-278.

Warren, D.A., T.G. Reigle, S. Muralidhara, and C.E. Dallas. 1996. Schedule-controlled operant behavior of rats following oral administration of perchloroethylene: Time course and relationship to blood and brain solvent levels. J. Toxicol. Environ. Health 47(4):345-362.

Werner, M., G. Birner, and W. Dekant. 1996. Sulfoxidation of mercapturic acids derived from tri- and tetrachloroethene by cytochromes P450 3A: A bioactivation reaction in addition to deacetylation and cysteine conjugate beta-lyase mediated cleavage. Chem. Res. Toxicol. 9(1):41-49.

White, I.N., N. Razvi, A.H. Gibbs, A.M. Davies, M. Manno, C. Zaccaro, F. De Matteis, A. Pahler, and W. Dekant. 2001. Neoantigen formation and clastogenic action of

HCFC-123 and perchloroethylene in human MCL-5 cells. Toxicol. Lett. 124(1-3):129-138.

WHO (World Health Organisation). 2008. Harmonization Project Document No. 6, Part 1: Guidance Document on Characterizing and Communicating Uncertainty in Exposure Assessment, International Program on Chemical Safety, World Health Organization, and Co-sponsored by International Labour Organization, and the United Nations Environmental Programme. Geneva, Switzerland: WHO [online]. Available: http://www.who.int/ipcs/publications/methods/harmonization/exposure_assessment.pdf [accessed Oct. 14, 2009].

Windham, G.C., D. Shusterman, S.H. Swan, L. Fenster, and B. Eskenazi. 1991. Exposure to organic solvents and adverse pregnancy outcome. Am. J. Ind. Med. 20(2):241-259.

Yoshioka, T., J.A. Krauser, and F.P. Guengerich. 2002. Tetrachloroethylene oxide: Hydrolytic products and reactions with phosphate and lysine. Chem. Res. Toxicol. 15(8):1096-1105.

Appendix A

Biographic Information on the Committee on Tetrachloroethylene

Sam Kacew (*Chair*) is a professor in the Department of Cellular and Molecular Medicine, Faculty of Medicine, and associate director of toxicology at the McLaughlin Centre for Population Health Risk Assessment of the University of Ottawa. His general research interests are in renal, hepatic, and pulmonary toxicology. Recent work has focused on the effects in infants of chemical contaminants in breast milk, the basis of differences between infants and children in responsiveness to chemicals, and the role of confounding factors in toxicity testing. Dr. Kacew is the recipient of several awards for his research and teaching. Most recently, he was awarded the Public Communications Award from the Society of Toxicology for his contribution to broadening public awareness of toxicologic issues through communication in books and public presentations. He is the editor-in chief of the *Journal of Toxicology and Environmental Health*, editor of the *Encyclopedia of Environmental Health*, and North American editor of *Toxicology and Environmental Chemistry*. He has served on numerous scientific expert panels and committees, including service as chair of the National Research Council Committee on Iodotrifluoromethane and member of the Committees on Depleted Uranium, Flame Retardants, and Jet Propulsion Fuel 8. He received his PhD in pharmacology from the University of Ottawa.

Bruce H. Alexander is an associate professor in the Division of Environmental Health Sciences of the University of Minnesota School of Public Health. His research interests are in applied occupational and environmental epidemiology, epidemiologic methods, and global health. Current research includes respiratory health and community exposure to asbestos-contaminated vermiculite; mortality, cancer incidence, and respiratory health in taconite production workers; health effects of occupational exposure to fluorochemicals; health effects of ionizing radiation in the medical field; pesticide exposure assessment in farm families; and the use of biologic markers in epidemiologic research. Dr. Alexander received his MS in environmental health from Colorado State University and his PhD in epidemiology from the University of Washington.

Margit Bleecker is director of the Center for Occupational and Environmental Neurology in Baltimore, Maryland. Her research interests are in clinical industrial neurotoxicology and occupational neurology. She has served on several Institute of Medicine committees, including two terms on the Committee to Review the Health Effects of Vietnam Veterans of Exposure to Herbicides. She received her PhD from the State University of New York Downstate Medical Center and her MD from the University of California, San Francisco School of Medicine. Dr. Bleecker is certified by the American Board of Psychiatry and Neurology.

Gary P. Carlson is professor of toxicology in the School of Health Sciences of Purdue University. His research interests are in examining the relationship between the metabolism of chemicals and their toxic actions, including an interest in activation and detoxification pathways in the liver and other target organs. Current research involves using a variety of techniques, ranging from in vitro assays to animal bioassays, to examine the biochemical mechanisms by which chemical agents exert their toxic and carcinogenic actions. He has served on several National Research Council committees, most recently as chair of the Subcommittee on Toxicologic Assessment of Low-Level Exposures to Chemical Warfare Agents and currently as a member of the Committee on Toxicology and the Committee on Combined Exposures to Hydrogen Cyanide and Hydrogen Monoxide in Army Operations. Dr. Carlson received his PhD in pharmacology from the University of Chicago.

Linda D. Cowan is George Lynn Cross Professor and chair of the Department of Biostatistics and Epidemiology of the University of Oklahoma Health Sciences Center. Her research interests include cardiovascular-disease epidemiology and the relative importance of risk factors in American Indian men and women, neurologic disorders, and perinatal epidemiology. Her recent research includes analysis of risk-factor profiles for early-onset and late-onset coronary heart disease in American Indians, investigation of the role of environmental toxicants and congenital hearing loss, and studies in west Africa of the prevalence of and risk factors for epilepsy associated with neurocysticercosis. Dr. Cowan has served on the National Research Council Committee to Assess the Health Implications of Perchlorate Ingestion and the Institute of Medicine (IOM) Committee to Assess the Safety and Efficacy of the Anthrax Vaccine. She is a member of the IOM Board on the Health of Select Populations. She received her PhD in epidemiology from Johns Hopkins University.

Mary E. Davis is a professor in the Department of Physiology and Pharmacology of the West Virginia University Health Sciences Center. Her research interests are in the toxicology of environmental and occupational pollutants, including water-disinfection byproducts, halogenated solvents, and arsenic. She is particularly interested in mechanisms of toxicity in the liver, kidneys, and vascular system. Dr. Davis was treasurer of the Society of Toxicology and is a former president of the society's Allegheny-Erie Regional Chapter. She has served on the U.S. Environmental Protection Agency Science Advisory Board and the editorial boards of *Toxicology* and *Toxicology and Applied Pharmacology*. She was a member of the National Research Council Committee on Assessing Human Health Risks of Trichloroethylene. She received her PhD in pharmacology from Michigan State University.

H. Christopher Frey is a professor in the Department of Civil, Construction, and Environmental Engineering of North Carolina State University. His research interests are in energy and environmental systems, specifically the development and application of methods for quantifying variability and uncertainty and for sensitivity analysis in system models. He has also been involved in exposure and risk analysis, particularly with regard to criteria pollutants, hazardous air pollutants, and particulate matter. Dr. Frey is a former president of the Society for Risk Analysis. He received his MS in mechanical engineering and his PhD in engineering and public policy from Carnegie Mellon University.

Joseph R. Landolph, Jr. is an associate professor of molecular microbiology and immunology and pathology at the Keck School of Medicine of the University of Southern California (USC). He also holds an appointment as associate professor of molecular pharmacology and pharmaceutical sciences in the USC School of Pharmacy. His research interests are in the molecular biology of chemical carcinogenesis induced by nickel and chromium compounds, specifically the processes of oncogene activation and tumor-suppressor gene inactivation in chemically induced neoplastic cell transformation. Other chemicals studied include carcinogenic arsenic compounds and polycyclic aromatic hydrocarbons. Dr. Landolph has held a number of leadership positions in the Society of Toxicology; he has been vice-president, president, and councillor of the Metals Specialty Section and councillor of the Carcinogenesis Specialty Section. He has previously served as a member of the U.S. Environmental Protection Agency (EPA) Human Health Strategies Review Committee and is a member of the Science and Technology Achievement Awards Committee and of the Drinking Water Committee of the EPA's Scientific Advisory Panel. He has served as a member of the Human Health Strategies Review Subcommittee of EPA's Board of Scientific Counselors. He received his PhD in chemistry from the University of California, Berkeley.

David C. McMillan is an associate professor in the Department of Cell and Molecular Pharmacology of the Medical University of South Carolina. His research interests are in the toxicity of drugs and environmental chemicals in erythrocytes and the liver. Current research is directed toward understanding the mechanisms underlying hemolytic anemia induced by drugs. Another line of research is aimed at understanding the role of metabolism in the carcinogenicity of trichloroethylene, specifically how known genetic variation in enzymes responsible for trichloroethylene metabolism alters the rates of production and the amounts of carcinogenic metabolites produced after exposure. Dr. McMillan received his PhD in pharmacology and toxicology from the University of Arkansas for Medical Sciences, and he is a diplomate of the American Board of Toxicology.

M.E. (Bette) Meek is the associate director of chemical risk assessment at the McLaughlin Centre of the University of Ottawa on interchange from Health Canada, where she managed the Existing Substances Division of the Safe Environments Programme of Health Canada. Her research interests are in hazard and risk assessment of chemical contaminants in the general environment. She led the development of approaches to establishing priorities for health assessment among the 23, 000 substances on the Canadian Domestic Substances List and approaches to in-depth risk assessment of high-priority substances. That included the introduction of novel predictive methods for exposure and hazard characterization, multimedia exposure estimation, chemical-specific adjustment factors for nonneoplastic effects, measures of potency for carcinogens, and robust models of peer engagement. More recently, she has been involved in the development of weight-of-evidence frameworks for mode of action based on consideration of mechanistic data in risk assessment. Dr. Meek has served as an adviser in those and related subjects to international scientific organizations (including the World Health Organization, the Organisation for Economic Cooperation and Development, the International Life Sciences Institute, and the International Labour Organization). She received her MSc in toxicology from the University of Surrey in the United Kingdom and her PhD in risk-assessment sciences at the University of Utrecht in the Netherlands.

M. Christopher Newland is Alumni Professor in the Department of Psychology of Auburn University. His research interests include the neurobehavioral toxicity of heavy metals, specifically the neurotoxicity of methylmercury during early development and aging, and behavioral pharmacology. He has served on advisory panels for the U.S. Environmental Protection Agency (EPA), the Agency for Toxic Substances and Disease Registry, and the National Research Council, where he has participated in reviews of the Neurotoxicology Division of the EPA Health Effects Laboratories and the neurotoxicity of elemental mercury, methylmercury, and manganese. His research has been supported by the National Institute of Environmental Health Sciences, National Institute on Drug Abuse, and EPA. He was a member of the Neurotoxicology and Alcohol Scientific Review Group. Dr. Newland is past president of the Neurotoxicology Specialty Section of the Society of Toxicology and past president of the Behavioral Toxicology Society. He has served on several editorial boards and is associate editor of *Neurotoxicology*. He received his MS and PhD in experimental psychology from the Georgia Institute of Technology and had postdoctoral fellowships in environmental health sciences (now environmental medicine) at the University of Rochester.

Julia B. Quint a research scientist, retired as chief of the Hazard Evaluation System and Information Service, Occupational Health Branch, of the California Department of Public Health. She was involved in identifying and evaluating reproductive toxicants, carcinogens, and other workplace chemical hazards and in developing strategies to protect workers, communities, and the environment from the hazards of toxic chemicals. Dr. Quint is a member of the California Environmental Contaminant Biomonitoring Program Scientific Guidance Panel and on the California Division of Occupational Safety and Health's Health Expert Advisory Committee for the Development of Permissible Exposure Limits for Airborne Contaminants in the Workplace. She received her PhD in biochemistry from the University of Southern California.

Gary L. Rosner is a professor of biostatistics at the University of Texas MD Anderson Cancer Center. He also holds an adjunct professorship in the Department of Statistics at Rice University and is a member of the faculty of the University of Texas at Houston Graduate School of Biomedical Sciences. His research interests are in population pharmacokinetics, pharmacodynamic modeling, pharmacogenetics, clinical-trial design, and Bayesian methods. Dr. Rosner has developed methods for analyzing complex biomedical data, such as those arising from population-based studies of the pharmacokinetics and pharmacodynamics of anticancer agents. He received his master's in applied mathetical sciences in applied mathematical sciences from Rice University and his ScD in biostatistics from Harvard University.

Ivan Rusyn is an associate professor in the Department of Environmental Sciences and Engineering of the University of North Carolina at Chapel Hill and associate director of the curriculum in toxicology. His research involves applying molecular, biochemical, genetic, genomic, and computational approaches to the understanding of the mechanisms of environmental-agent-related organ injury and carcinogenesis. Recent work has focused on the molecular mechanisms of phthalate-induced carcinogenesis, mechanisms of ethanol-induced hepatic toxicity based on the latest knowledge of the genetic diversity of the mouse as a model organism, and genomic and genetic analysis of hepatic and renal toxicity of trichloroethylene to determine what genetic variants correlate with susceptibility or resistance to hepatic disease. Dr. Rusyn received his MD from the Ukrainian State Medical University in Kiev and his PhD in toxicology from the University of North Carolina at Chapel Hill.

Rolf Schulte-Hermann is emeritus professor of toxicology at the Medical University of Vienna. He was head of the Research Unit Chemical Safety and Cancer Prevention, and, from 1985 to 2004, director of the Institute of Cancer Research at the University of Vienna. His research interests are focused on regulation of organ growth, tumor initiation and promotion, non-genotoxic carcinogens, and role of the microenvironment in chemical carcinogenesis. Major scientific achievements include the discovery of apoptotic cell death during organ regression and carcinogenesis and of apoptosis inhibition by tumor promoters. In 1991 he founded the Austrian Society of Toxicology and served as chairman until 2009. He is director of the Postgraduate Course in Toxicology/Chemical Safety in Vienna since 1993. He served as member of numerous national and international advisory committees. Dr. Schulte-Hermann received his PhD in pharmacy from the Free University Berlin and his MD from the University of Marburg, Germany.

Irvin R. Schultz is a toxicologist in the Marine Sciences Laboratory of Pacific Northwest National Laboratory, operated by Battelle for the U.S. Department of Energy in Sequim, Washington. He also holds an appointment as an adjunct assistant professor in the Department of Biology of the University of Idaho. His research interests cover both ecologic and human health issues. Highlights of his research efforts include studies of the disposition of drinking-water disinfection byproducts in human volunteers, nonhuman primates, and rodent models; development of physiologically based toxicokinetic models to describe the chemical dosimetry and estrogenic activity of xenobiotics; the metabolism and disposition of environmental pollutants in fish, with an emphasis on allometric and interspecies scaling; and the disposition and bioavailability of inorganic and organometallic compounds in fish. Dr. Schultz received his PhD in pharmacology-toxicology from Washington State University.

Robert Snyder is associate dean for research of the Ernest Mario School of Pharmacy of Rutgers University and was a professor and chair of the Department of Pharmacology and Toxicology of Rutgers College of Pharmacy, director of the of the Environmental and Occupational Health Sciences Institute, director of its Division of Toxicology, and director of the Graduate Program in Toxicology. His research interests are in solvent toxicology, chemically induced bone marrow depression, hepatic toxicity, chemical carcinogenesis, and drug metabolism. He has done extensive work on benzene leukemogenesis. Dr. Snyder is a former president of the American College of Toxicology and has served on several committees of the National Research Council, most recently on the Committee for Review and Assessment of the Army Non-Stockpile Demiltarization Program: Workplace Monitoring.

Roberta F. White is a professor and chair of the Department of Environmental Health of the Boston University School of Public Health. She also is associate dean of research and holds appointments in the Department of Neurology and the Department of Psychology of the university. Her research interests are in the effects of exposures to industrial chemicals and chemical pollutants on brain function on the basis of behavioral measures and neuroimaging techniques. She has studied behavioral and imaging correlates of occupational lead exposure and environmental exposure to methylmercury, structure-function relationships revealed by visuospatial tests, solvent exposures of children and adults, and effects of prenatal pesticide exposure in farm workers in South Africa. Dr. White received her PhD in clinical psychology from Wayne State University.

Luoping Zhang is an associate adjunct professor in the Division of Environmental Health Sciences of the University of California, Berkeley. Her research interests are in mechanisms of bone marrow toxicity caused by benzene and other toxic chemicals, application of fluorescent in situ hybridization as a biomarker in studies of childhood leukemia and other types of cancer, and application of gene-expression profiling in molecular epidemiology. She received her MS in biochemistry from Huazhong University of Science and Technology in the People's Republic of China and her PhD in biochemical toxicology from Simon Fraser University, in British Columbia, Canada.

Yiliang Zhu is a professor in the Department of Epidemiology and Biostatistics of the University of South Florida College of Public Health and director of the college's Center for Collaborative Research. His current research is focused on quantitative methods in health risk assessment, including physiologically based pharmacokinetic models, dose-response modeling, benchmark-dose methods, and uncertainty quantification. He also conducts research in disease surveillance, health-outcome evaluation, and health-care access and use in developing countries. Dr. Zhu was a member of the National Research Council Committee on EPA's Exposure and Human Health Assessment of Dioxin and Related Compounds. He received his MS in statistics from Queen's University and his PhD in statistics from the University of Toronto.

Appendix B

Dissenting Statement and Rebuttal

Dissenting Statement on Mode of Action of Tetrachloroethylene in Mouse Hepatocarcinogenesis

By Rolf Schulte-Hermann

The authors of the Integrated Risk Information System (IRIS) draft conclude in Chapter 4.4.

- That peroxisome proliferator-activated receptor-alpha (PPARα) activation is not the primary mode of action (MOA) for tetrachloroethylene-induced hepatocarcinogenesis in mice.
- That the specific mechanisms or MOAs for hepatocarcinogenesis are not known.
- That it is highly likely that more than one mechanism is operative.

That conclusion is supported in Chapter 6 of the present committee review of the IRIS draft although some deficiencies in the draft are mentioned. They include the lack of coherent flow and an imbalance in critiquing the view that the PPARα MOA is not relevant for human carcinogenesis. This committee member concurs with the criticisms.

However, the member disagrees with the conclusions quoted above. In the members' opinion, the weight of evidence strongly favors a key role of PPARα activation in tetrachloroethylene-induced hepatocarcinogenesis in mice; furthermore, this MOA lacks relevance for human hepatocarcinogenesis. Because of the deficits in the respective presentation in the IRIS draft, the following paragraphs will briefly compile the essential data supporting the PPARα MOA for tetrachloroethylene, the role of trichloroacetic acid (TCA) as the major responsible metabolite of tetrachloroethylene, the potential roles of other MOAs, new mechanistic data supporting the lack of relevance of the PPARα MOA for humans. The author hopes that the arguments collected in this dissent will be helpful in revising the IRIS draft.

EVIDENCE THAT TETRACHLOROETHYLENE AND TRICHLOROACETIC ACID ARE PEROXISOME PROLIFERATORS

Relevance of Trichloroacetic Acid vs Dichloroacetic Acid

Both TCA and dichloroaceticacid (DCA) are peroxisome proliferators. TCA is the major metabolite found in the body after exposure to tetrachloroethylene. It is eliminated slowly and therefore accumulates to some extent. In contrast, DCA is present in only tiny amounts after tetrachloroethylene exposure because of low formation and more rapid elimination (IRIS draft, Chapter 3). Thus, after tetrachloroethylene administration in mice, DCA concentrations in blood were below 10 or 25 µg/mL in the initial hours and then undetectable and were undetectable in the liver in the presence of high TCA concentrations (up to 150 µg/mL or 150 µg/g) (Philip et al. 2007; see below for experimental details). TCA and DCA have similar potency as hepatic carcinogens and tumor promoters (Bull 2000; Bull et al. 2004). Overall, therefore, DCA probably contributes little to PPARα-mediated effects of tetrachloroethylene. Other metabolites of tetrachloroethylene are not known to be peroxisome proliferators. The arguments related to the PPARα MOA should therefore focus on TCA.

Peroxisome Proliferator-Activated Receptor-Alpha Transactivation

Tetrachloroethylene (up to 5 mM) did not transactivate mouse and human PPARα in cells transfected with the PPAR genes. Likewise, chloral hydrate and trichloroethanol, minor metabolites of tetrachloroethylene, did not activate PPARα. In contrast, TCA was active at 1 and 5 mM but not at 0.1 mM. Activity was considerable at 1 mM, suggesting that the lowest observed-adverse-effect level (LOAEL) for binding activity is distinctly below 1 mM (Zhou and Waxman 1998; Maloney and Waxman 1999). The maximal activation was only about 50% of that of Wy 14643, a strong activator, but similar to that of mono-(2-ethylhexyl) phthalate, the carcinogenic metabolite of di(2-ethylhexyl) phthalate (DEHP). Mouse PPARγ displayed little, and human PPARγ no, responsiveness to TCA. DCA transactivated PPARα with somewhat less potency than TCA, but it showed no effect on mouse or human PPARγ (Zhou and Waxman 1998; Maloney and Waxman 1999). In another study (Walgren et al. 2000), TCA but not DCA was found to activate mouse PPARα at 4 mM.

Tetrachloroethylene as a Peroxisome Proliferator

Tetrachloroethylene induces in mouse liver responses that are known to be mediated by PPARα—such as a 4-fold increase in CN-insensitive palmitoyl-CoA oxidation (PCO), morphologic evidence of peroxisome proliferation based on morphometric analysis, and hepatomegaly—at doses of 1,000 mg/kg by gavage for 10 days or 200 and 400 ppm by inhalation 6 hours/day for 14, 21, or 28 days (Goldsworthy and Popp 1987; Odum et al. 1988). Those effects also occurred, although much more weakly, in rats (Goldsworthy and Popp 1987; Odum et al. 1988). Dose-dependent increases in hepatomegaly and (not significantly) hepatocyte proliferation after oral treatment of mice were reported by Schumann et al. (1980) and Buben and O'Flaherty (1985). In male mice, tetrachloroethylene at daily oral doses of 150, 500, and 1,000 mg/kg transiently and dose-dependently increased hepatocyte DNA synthesis at 7 and 14 days; at 30 days, the increase was nearly gone (Philip et al. 2007). Tetrachloroethylene itself does not bind to PPARα (see above), so PPARα-mediated responses should be due to active metabolites, predominantly TCA. In the study by Odum (1988), 200 ppm, the higher dose in the NTP (1986) carcinogenicity study, induced pronounced increases in PCO and peroxisome proliferation that suggested that the NOAEL was much lower. Obviously, doses of tetrachloroethylene that are in the range of the carcinogenic doses are transformed to metabolites (mainly TCA) in amounts sufficient to activate PPARα in mouse liver. Evidence supporting the role of TCA is presented later.

Trichloroacetic Acid as a Peroxisome Proliferator and Hepatocarcinogen in Mice

TCA was shown in numerous studies to induce PPARα-mediated responses, such as PCO increases, in the livers of mice of both sexes and to produce liver tumors in mice (Goldsworthy and Popp 1987; Pereira 1996; Bull 2000; Bull et al. 2002; DeAngelo et al. 2008; further references in the IRIS draft). In the first days of administration, TCA induced liver enlargement and an increase in hepatocyte DNA synthesis in male and female mice (Dees and Travis 1994; Pereira 1996; Stauber and Bull 1997). Effects were present when TCA was given at 100 mg/kg orally over 11 days and showed some increase with dose up to 1,000 mg/kg (Dees and Travis 1994). With continued treatment, the enhancement of DNA synthesis disappeared and was reversed to depression (Pereira 1996; Stauber and Bull 1997). TCA induction of the peroxisomal enzymes PCO and acyl-CoA oxidase (by RNA expression) and of CYP 4a depended on the presence of the PPARα gene and were not seen in PPARα-null mice (Laughter et al. 2004). Some studies reported increased lipid peroxidation by TCA (Bull et al. 1990; Larson and Bull 1992; Austin et al. 1996). An increase in 8-OHdG was not found after TCA (Parrish et al. 1996) or after tetrachloroethylene (Toraason et al. 1999).

Hepatic tumorigenesis after TCA administration was studied mostly in male mice but was also demonstrated in female mice (Pereira 1996). TCA was found to promote hepatic-tumor development efficiently in mice after initiation by 1-methyl-1-nitrosourea or vinyl carbamate (Pereira and Phelps 1996; Bull et al. 2004). Foci of altered cells (presumably preneoplastic lesions) and tumors were predominantly basophilic and did not express glutathione *S*-transferase-pi (GSTP), as found with other peroxisome proliferators (Pereira 1996; Pereira and Phelps 1996; Stauber and Bull 1997). Clonal expansion of anchorage-independent hepatocytes obtained from male B6C3F1mice by administration of TCA in vitro was also reported (Stauber et al. 1998).

In a recent lifetime dose-response study, DeAngelo et al. (2008) found that the TCA-induced increase in PCO correlated with tumor induction, and a linear association occurred between the two effects. A TCA NOAEL of 6 mg/kg per day and a LOAEL of 58-68 mg/kg were reported.

Evaluation of Effects of Trichloroacetic AcidTCA and Tetrachloroethylene for Consistency with Key Events

Klaunig et al. (2003) have defined seven key events in the PPARα MOA of rodent hepatocarcinogenesis. TCA was found to induce most of the key events in mice:

1. **Causal relationship** to tumor formation:
   a. Direct activation of PPARα (resistance to induction of key events in PPARα-null mice).
   b. Transient increase in hepatocyte DNA synthesis.
   c. Selective clonal expansion of the putative preneoplastic lesions and of tumors.
2. **Associative relationship** to tumor formation:
   a. Peroxisome proliferation as indicated by morphologic and biochemical studies (high weight of evidence and specificity for association with tumorigenesis [Klaunig et al. 2003]).
   b. Hepatocyte oxidative stress (lipid peroxidation) (low weight of evidence and specificity for association [Klaunig et al. 2003]).
   c. Inhibition of gap junctional intercellular communication (GJIC) by TCA in a model with lucifer yellow. The same result was obtained with tetrachloroethylene (Benane et al. 1996)
   d. Dependence on Kupffer cells has apparently not been studied directly after TCA. administration. However, that is not a serious deficiency for the purpose of this discussion, because the specificity of Kupffer-cell dependence is low (Klaunig et al. 2003).

This set of results was generated in several studies, and dose-response and temporal relationships are consistent with the observation of tumors. In the absence of evidence on genotoxicity and other plausible MOAs, the induction of 6 of the 7 key events provide strong evidence of a PPARα-dependent MOA of TCA-induced mouse hepatocarcinogenesis. The same conclusion was reached by the National Research Council's Committee on Human Health Risks of Trichloroethylene (2006).

Data on tetrachloroethylene are less comprehensive. An NOAEL and a LOAEL and studies in PPARα-null mice are not available. Nevertheless, the PPARα MOA is considered probable on the basis of the induction of several key events in mouse liver, including transient increases in DNA synthesis, lipid peroxidation, inhibition of GJIC, and, most important, peroxisome proliferation, an event highly specific for PPARα activation. A major argument supporting the PPARα MOA of tetrachloroethylene is related to the role of TCA as the active metabolite, as will be shown below according to several lines of evidence.

SPECIES DIFFERENCES SUPPORTING THE ROLE OF TRICHLOROACETIC ACID AS THE ACTIVE METABOLITE OF TETRACHLOROETHYLENE

Rats are less sensitive than mice to peroxisome-proliferator effects of the same doses of tetrachloroethylene (Goldsworthy and Popp 1987; Odum et al. 1988) and do not develop hepatic tumors in response to it (NCI 1977; NTP 1986; JISA 1993) or to TCA at doses up to 364 mg/kg per day for 104 weeks (DeAngelo et al. 1989, 1997). Those differences can be explained by the kinetics of tetrachloroethylene in the two species. Mice metabolize the agent and form TCA at concentrations several times higher than do rats (Schumann et al. 1980; Reitz et al. 1996). Thus, the area under the curve (AUC) for blood TCA after exposure to tetrachloroethylene at 400 ppm for 6 hours was 6.7 times higher in mice than in rats (Odum et al. 1988). In addition, mice are more sensitive than rats to induction of peroxisome proliferation by TCA. That may, at least partially, be due to the 10-fold higher binding capacity of rats' than of mice's plasma proteins for TCA (maximal binding capacity, 283 μM in rats and 29 μM in mice). As a result, the proportion of TCA available for uptake by the liver will be less in rats than in mice and will produce a weaker response in rats (Lumpkin et al. 2003). The weak peroxisome-proliferator effect seen in rats is obviously insufficient for hepatic-tumor formation. Numerous examples show that low levels of induction of peroxisomes are not necessarily associated with hepatic tumorigenesis (Klaunig et al. 2003). Overall, the striking differences between responses of mice and of rats to tetrachloroethylene can be explained by assuming TCA as the active principle.

CARCINOGENICITY STUDIES WITH TETRACHLOROETHYLENE AND TRICHLOROACETIC ACID

Hepatocarcinogenic doses of tetrachloroethylene in mice are displayed in

Tables 1 and 2. Doses that do not increase rates of hepatocarcinoma were not tested in National Cancer Institute (NCI) and National Toxicology Program (NTP) studies. Rats treated in parallel bioassays did not develop hepatic tumors.

Long-term exposure to TCA was shown to result in hepatic-tumor formation in mice (Table 3A) but not rats (DeAngelo et al. 1997). DeAngelo et al. (2008) exposed male B6C3F1 mice to TCA via drinking water at 0.05, 0.5, 4.5, and 5 g/L for 60 and 104 weeks (Table 3B). Daily doses calculated were 6-8, 58-68, and 572-602 mg/kg. The work consisted of three parts conducted in two Environmental Protection Agency (EPA) laboratories. The authors reported significant increases in the prevalence and multiplicity of hepatic tumors in the two higher dose groups. A TCA NOAEL of 6 mg/kg per day and a LOAEL of 58-68 mg/kg per day were derived for neoplastic and nonproliferative pathology.

Somewhat surprisingly, the IRIS draft does not mention parts 1 and 2 of the study by DeAngelo et al. 2008), which is therefore presented completely in Table 3B. The selection of only one of the two 104-week bioassays has important implications for modeling in Appendix 4A of the IRIS draft because the control group selected shows a dramatically higher hepatic-tumor incidence (64% vs 12%; see parts 3 and 2 in Table 3B). Use of the low-tumor control would increase the fraction of animals affected by TCA (Figure 4A-1 of the IRIS draft) and increase the calculated carcinogenic potency of TCA. To add to the confusion, in the publication of DeAngelo et al. (2008), the allocation of controls and treated groups in parts 2 and 3 of the study is at variance between the methods section and Table 6 of the results section. That discrepancy should be resolved, and all pertinent data should be used in revising the IRIS document. At present, the validity of the modeled TCA potency data as used in Appendix 4A is questionable.

TABLE 1 Carcinogenicity Study in B6C3F1 Mice (Tetrachloroethylene In Corn Oil Was Administered By Gavage 5 Time a Week for 78 Weeks and Followed By an Observation Period of 12 Weeks)

Sex	Bioassay	Dose, mg/kg (TWA)	Carcinoma (Incidence)	Mice at Risk
Male	NCI	0	2	17
		0 (vehicle)	2	20
		536	32	49
		1,072	27	48
Female	NCI	0	2	20
		0 (vehicle)	0	20
		386	19	48
		772	19	48

Source: NCI 1977.

TABLE 2 Incidence of Hepatocellular Adenomas and Carcinomas in B6C3F1 Mice Exposed to Tetrachloroethylene in Two Inhalation Bioassays

			Cumulative LiverTumors at Week 104			
Sex	Bioassay	Administered Exposures, ppm	Adenomas	Carcinomas	Adenomas or Carcinomas	Total at Risk[a]
Male	NTP (1986)	0	12	7	17	49
		100	8	25	31	47
		200	19	26	41	50
	JISA (1993)	0	7	7	13	46
		10	13	8	21	49
		50	8	12	19	48
		250	26	25	40	49
Female	NTP (1986)	0	3	1	4	45
		100	6	13	17	42
		200	2	36	38	48
	JISA (1993)	0	3	0	3	50
		10	3	0	3	47
		50	7	0	7	48
		250	26	14	33	49

[a]Animals that died before the first appearance of a hepatocellular tumor, but no later than week 52, were omitted from the totals because they were presumed not to have adequate time in the study to develop tumors.
Source: EPA 2008 (Table 4A-3).

TISSUE CONCENTRATIONS OF TRICHLOROACETIC ACID AFTER ADMINISTRATION OF TETRACHLOROETHYLENE OR TRICHLOROACETIC ACID

A key question in identification of the MOA of tetrachloroethylene-induced hepatic tumors is whether sufficient TCA is formed and available in the target organ for effective induction of peroxisome proliferation and hepatocarcinogenesis. To address that question, a literature search has been conducted for analytic data on TCA concentrations in the liver and, as a surrogate, in the blood. The results are described below and displayed in Tables 4A-D and Figures 1 and 2A-C.

Blood and Liver Concentrations of Trichloroacetic Acid After Administration of Tetrachloroethylene

Blood concentrations of TCA after tetrachloroethylene administration were first analyzed by Odum et al. (1988). After a single exposure at 400 ppm for 6 hours, peak blood concentrations in B6C3F1 mice were 130 μg/mL 3-4 hours after the end of exposure and thereafter declined with a half-life of 7-8 hours. The AUC was calculated (Table 4A).

TABLE 3A Trichloroacetic Acid Drinking-Water Studies in Male Mice: Incidence of Hepatocellular Adenomas and Carcinomas

Source	Weeks of Exposure	TCA Exposure, g/L	Equivalent TCA Exposure, mg/kg-day	N	Incidence of Adenomas	Incidence of Carcinomas	Incidence of Adenomas or Carcinomas	Proportion Responding with Carcinomas
Bull et al. (1990)[a]	37	2	330	11	0	3	3	0.27
	52	0	0	35	0	0	0	0.0
		1	170	11	2	2	NR	0.18
		2	330	24	1	4	NR	0.17
Bulll et al. (2002)	52	0	0	20	0	0	0	0.0
		0.5	NR	20	5	3	6	0.15
		2	NR	20	6	3	8	0.15
Herren-Freund et al. (1987)	61	0	0	22	2	0	2	0.0
		5	NR	22	8	7	NR	0.32
Ferreira-Gonzalez et al. (1995)	104	0	0	16[b]	NR	3[b]	NR	0.19
		4.5	NR	11	NR	8	NR	0.73
DeAngelo et al. (2008)	104	0	0	56	10	26	31	0.55
		0.05	8	48	10	14	21	0.44
		0.5	68	51	20	32	36	0.71

[a]Cumulative TCA exposures were provided in grams per kilogram for the mice evaluated at 52 weeks. Those exposures were converted to milligrams per kilogram per day by (1,000 mg/g)/(7 days/[week][52 weeks]).
[b]Estimated from the reported proportion responding by selecting the smallest group size and incidence value consistent with the precision of the reported proportion.
NR = not reported.
Source: EPA 2008 (Table 4A-1).

TABLE 3B Complete Presentation of Results of the TCA Carcinogenicity Study of DeAngelo et al. (2008)

Weeks	TCA, g/L	Equivalent TCA, mg/kg	N[a]	Number with Denoma	Number with Carcinoma	Number with Adenoma or Carcinoma
60 (part1)	0 (NaCl)	0	30	7	7	13
	0.05	8	27	15	4	15
	0.5	68	29	21	21	38
	5.0	602	29	38	38	55
104 (part 2)	0 (NaCl)	0	25	0	12	12
	4.5	572	36	59	78	89
104 (part 3)	0 (1.5 g of acetic acid)	0	42	21	55	64
	0.05	6	35	23	40	57
	0.5	81	37	51	78	87

[a]Number of animals examined.

Note: Table 3A from EPA (2008) contains only part 3 and reports higher numbers of animals examined than the publication by De Angelo et al. and somewhat different proportions of carcinomas.

TABLE 4A Blood TCA Concentrations After Tetrachloroethylene Treatment

1) 1x 400 ppm for 6 hours	Odum et al. 1988				
Doses (ppm x 6 hours)	400				
Peak concentrations (µg/mL)[a]	130				
AUC_{0-24} (µg/mL per hour)[b]	1,760				
2) 1x i.g.	Gearhart et al. 1993			Ratios	
	1	2	3	1 - 2	2 - 3
Doses (mg/kg)	100	536	1,072	5.36	2.0
Peak concentrations (µg/mL)[a]	23	80	157	3.48	1.96
AUC_{0-24} (µg/mL per hour)[b]	368	1,317	2,840	3.58	2.16
3) 1x, i.g.; SW mice	Philip et al. 2007			Ratios	
	1	2	3	1 - 2	2 - 3
Doses (mg/kg)	150	500	1,000	3.33	2.0
Peak concentrations (µg/g)[a]	150	160	170	1.07	1.06
AUC_{0-24} (µg/mL per hour)[a]	2,583[c]	2,229	3,208	0.86	1.44
4) 30x, daily i.g.; SW mice	Philip et al. 2007			Ratios	
	1	2	3	1 - 2	2 - 3
Doses (mg/kg)	150	500	1,000	3.33	2.0
Peak concentrations (µg/g)[a]	75	128[c]	130	1.71	1.02
AUC_{0-24} (µg/mL per hour)[a]	864	2197[c]	2,439	2.54	1.11

[a]Numbers read from figure.

[b]Calculated from figure.

[c]Data of the first two time points were excluded from the calculation.

Note: If not indicated otherwise, male B6C3F1 mice were used. Ratios between doses, peak TCA concentrations, and AUC are indicated. i.g. = intragastric application. Further technical data on the studies is given in the text.

TABLE 4B Blood TCA Concentrations after TCA Treatment

1) 1x i.g., 4-hour fast	Larson and Bull 1992			Ratios	
	1	2	3	1 - 2	
Doses (mg/kg)		20	100	5	
C_{max} (μg/mL)		38 ± 1.65	130 ± 9.9	3.4	
AUC_{0-24} (μg/mL per hour)		333 ± 9.9	1185 ± 34.7	3.5	
2) 1x i.g., 8-hour fast	Templin et al. 1993			Ratios	
	1	2	3	1 - 2	2 - 3
Doses (mg/kg)	5	20	100	4	5
Peak concentrations (μg/mL)[a]	10.1	40.3	80.6	4.0	2.0
AUC_{0-24} (μg/mL per hour)[a]	87	374	934	4.3	2.5
3a) 1x i.v., 16-hour fast	Gonzalez-Leon 1999				
Doses (mg/kg)			100		
C_{max} (μg/mL)			179 ± 30		
AUC_{0-24} (μg/mL per hour)			2,516 ± 289		
3b) Pretreatment with TCA at 2 g/L for 14 days, then 1x i.v., 16-hour fast					
Doses (mg/kg)			100		
C_{max} (μg/mL)			214 ± 17		
AUC_{0-24} (μg/mL per hour)			2,964 ± 418		
4) Drinking water, 14 days	Mahle et al. 2001			Ratios	
	1	2	3	1 - 2	2 - 3
Doses (mg/kg)	11.6	110	268	9.5	2.4
Peak concentrations (μg/mL)	10.3	72.9	79.9	7.1	1.1
5) Drinking water for 5 or 14 days	Green 2003 (Data from Sweeney et al. 2009)			Ratios	
	1	2	3	1 - 2	2 – 3
Doses (mg/kg), 5 days		180	443		2.5
Peak concentrations (μg/mL)		71.6	127		1.8
Doses (mg/kg), 14 days		181	497		2.8
Peak concentrations (μg/mL)		97.5	133		1.4

[a]Numbers read from figure.
Note: For explanations, see Table 4A.

TABLE 4C Liver TCA Concentrations after Treatment with Tetrachloroethylene

1) 1x, i.g.; SW mice	Philip et al. 2007			Ratios	
	1	2	3	1 - 2	2 - 3
Doses (mg/kg)	150	500	1,000	3.33	2.0
Peak concentrations (μg/g) [a]	53	100	175	1.89	1.75
AUC_{0-24} (μg/mL per hour) [a]	956	1,690	3,233	1.77	1.91
1) 30x, daily i.g.; SW mice	Philip et al. 2007			Ratios	
	1	2	3	1 - 2	2 - 3
Doses (mg/kg)	150	500	1,000	3.33	2.0
Peak concentrations (μg/g) [a]	25	34	42	1.36	1.24
AUC_{0-24} (μg/mL per hour) [a]	388	563	694	1.45	1.23

[a]Numbers read from figure.
Note: For explanations, see Table 4A.

TABLE 4D Liver TCA Concentrations after Treatment with TCA

2) 1x i.g., 8-hour fast	Templin et al. 1993			Ratios	
	1	2	3	1 - 2	2 - 3
Doses (mg/kg)	5	20	100	4	5
Peak concentrations (μg/g)	6.4	21.1	28.4	3.3	1.3
AUC_{0-24} (μg/mL per hour)	55	199	386	3.6	1.9
4) Drinking water, 14 days	Mahle et al. 2001			Ratios	
	1	2	3	1 - 2	2 - 3
Doses (mg/kg)	11.6	110	268	9.5	2.4
Peak concentrations (μg/mL)	6.2	48.2	61.6	7.77	1.28

Note: For explanations, see Table 4A.

Gearhart et al. (1993) administered a single dose of tetrachloroethylene to male B6C3F1 mice by gavage in corn oil at of 0.1, 0.536, and 1.072 mg/kg. The two higher doses correspond to those used in the NCI oral-carcinogenicity study (Table 1). TCA reached peak blood concentrations of 23, 80, and 157 mg/l; these and the AUC are shown in Table 4A.

In a similar study of male Swiss Webster mice, Philip et al. (2007) applied tetrachloroethylene in aqueous gavage (with Emulphor) daily in three dosages (150, 500, and 1,000 mg/kg) for up to 30 days. Concentrations of tetrachloroethylene, TCA, DCA, and trichloroethanol were analyzed after one and 30 treatments. After the first treatment, peak blood TCA was similar with all three dosages. After 30 doses of tetrachloroethylene at 150 mg/kg, blood TCA ranged from 35 to 75 μg/mL in the 24-hour period, and after 500 and 1,000 mg/kg, from 50 to 135 μg/mL. Peak concentrations and the AUC are displayed in Table 4A. Peak hepatic TCA and AUC tended to be lower than the corresponding blood concentrations, particularly after 30 days of treatment (Table 4C).

Table 4 also shows ratios between different doses compared with ratios between the corresponding peak concentrations and AUC values. Although tissue concentrations in general increased with dose, the relative difference tended to decrease with increasing dose. That reflects the well-known fact that the metabolism of tetrachloroethylene is saturable (Buben and O`Flaherty 1985; Reitz et al. 1996).

Blood and Liver Trichloroacetic Acid Concentrations After Administration of Trichloroacetic Acid

TCA concentrations after administration of TCA in mice and rats have been measured in several studies. After a single oral dose of 20 or 100 mg/kg ^{14}C-TCA in male B6C3F1 mice, TCA C_{max} in blood were 38 and 130 µg/mL. Half-lifes ($T_{1/2}$) were 4.2 and 5.8 hours; for AUC data, see Table 4B (Larson and Bull 1992). In male B6C3F1 mice treated orally with TCA at single doses of 5, 20, and 100 mg/kg, peak blood concentrations were 10.1, 40.3, and 80.6 µg/mL, respectively, and the half-life was 5.4-6.4 hours. Liver concentrations—6.4, 21.1, and 28.4 µg/g—were lower than blood concentrations; it was suggested that this result from plasma-protein binding of TCA. For AUC data, see Tables 4B and D. Liver:blood AUC ratios decreased with increasing dose (Templin et al. 1993).

When given intravenously to male B6C3F1 mice, a "challenge dose" of TCA at 100 mg/kg resulted in a blood C_{max} of 179 µg/mL and $t_{½}$ was 10.0 hours. Other mice received TCA for 14 days at 2 g/L in drinking water. The challenge dose was then administered 16 hours after removal of TCA from drinking water. No significant changes in various kinetic measures occurred: blood C_{max} was 214 µg/mL, $t_{½}$, 9.4 hours; metabolism of TCA in vitro was not altered. The authors concluded that pretreatment with TCA does not affect metabolism and pharmacokinetics of TCA (Gonzalez-Leon et al 1999).

In a similar study, male B6C3F1 mice received TCA at 0.08, 0.8, or 2.0 g/L in drinking water; this resulted in daily dose rates of 11.6, 110, and 268 mg/kg. After 14 days, blood TCA was 10.3, 72.9, and 79.9 µg/mL; they were almost identical after 3 days. Liver TCA at 14 days was 6.2, 48.2, and 61.6 µg/mL (Tables 4B and D) (Mahle et al. 2001).

Available studies of organ TCA concentrations used male mice except that Green et al. (cited from Sweeney et al. 2009) found even lower blood concentrations in female B6C3F1 mice exposed to TCA than in male mice.

In conclusion, blood and liver concentrations after TCA treatment in different studies are fairly consistent at similar doses. Liver concentrations were lower than blood concentrations. The data from Larson and Bull (1992), Templin et al. (1993), Mahle et al. (2001), and Green et al. from CTL (cited by Sweeney et al. 2009) concordantly demonstrate that peak blood and liver TCA concentrations and AUC do not increase linearly with dose. Rather, as shown by the ratios in Tables 4B and D, the increments decreased with increasing dose.

Above 100 mg/kg, little further increase in peak blood concentrations is apparent from the experimental data available. That result is graphically presented in Figure 1. Obviously, bioavailability of TCA administered orally does not increase linearly with dose in mice. Incomplete bioavailability of oral TCA is independently supported by the study of Gonzalez-Leon et al. (1999), in which TCA was administered intravenously at 100 mg/kg. Peak concentrations and AUC were about 2.5 times higher than the mean in studies that used oral administration (Table 4B). Obviously, a large portion of a 100-mg/kg dose of TCA administered orally is not systemically available. Elimination kinetics in blood after various doses of TCA were similar and repeated treatment with TCA (for 14 days) did not significantly modify its metabolism and kinetics (Templin et al. 1993; Mahle et al. 2001; Gonzalez-Leon et al. 1999), so a dose-dependent limit on absorption of TCA seems a likely explanation of the reduced bioavailability of oral TCA.

The fraction of TCA bioavailable after oral exposure was modeled by Sweeney et al. (2009) on the basis of blood-concentration data of Mahle et al. (2001) and Green et al. They concluded that the apparent bioavailability of TCA from drinking water is 25% at low doses (12 mg/kg) and declines to less than 10% at high doses (800 mg/kg).

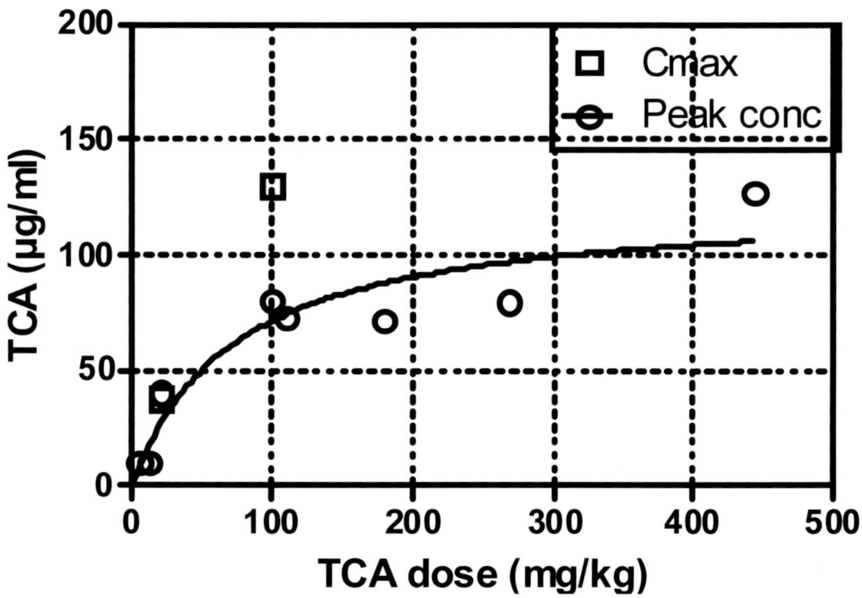

FIGURE 1 Peak TCA concentrations and C_{max} in blood after oral administration of TCA. Source: Data from Table 4B.

Conclusions on the Role of Trichloroacetic Acid in Tetrachloroethylene-Induced Hepatocarcinogenesis

Carcinogenic Potency

The validity of the modeled carcinogenic-potency data on TCA in the IRIS draft (Appendix 4A) is questionable, see earlier section on Carcinogenicity studies with tetrachloroethylene and TCA.

Direct Comparison of Trichloroacetic Acid Concentrations in Blood or Target Organ

Tables 4A and B display the available blood TCA concentrations as determined analytically. Peak TCA concentrations and AUC are similar after application of tetrachloroethylene and TCA at carcinogenic doses or perhaps even higher after tetrachloroethylene than after TCA. That point is illustrated by Figures 2A-C. Likewise, the corresponding liver TCA concentrations are similar after both agents or even higher after tetrachloroethylene. That is convincing evidence that TCA can be formed from tetrachloroethylene and be present in blood and target organ in amounts sufficient to induce peroxisome proliferation and hepatocarcinogenesis.

Modeling the Internal Trichloroacetic Acid Dose

In the IRIS draft, a quantitative comparison between hepatic-carcinoma yields after tetrachloroethylene or TCA treatment and the corresponding internal TCA doses is attempted. Internal TCA after tetrachloroethylene was modeled according to Reitz et al. (1996). For modeling internal TCA after TCA treatment, an absorption rate of 95% was estimated (Section 4A1.2, p.4-205). No reference or reason for that estimate is provided, and no support was found in the literature. Clearly the IRIS estimate is not compatible with the available literature reviewed above, which demonstrates that TCA absorption after oral exposure is incomplete and decreases with increasing dose (Tables 4B and D, Figure 1). Moreover, modeling by Sweeney et al. (2009) suggests that in the 10 mg/kgdose range only 25% of oral TCA becomes bioavailable. Apparently, therefore, the internal TCA doses after TCA treatment that are calculated in the IRIS draft are too high. In consequence, tumor yields predicted to result from TCA formed after tetrachloroethylene (Table 4A-4of the IRIS draft) would be too low. Indeed, when they included their bioavailability data in the model, Sweeney et al. (2009) found that TCA in mice exposed to tetrachloroethylene is sufficient to explain the incidence of hepatic tumors. In conclusion, formation of TCA from tetrachloroethylene is probably sufficient to explain tumorigenesis in mouse liver. That adds substantially to the weight of evidence of a key role of

PPARα activation in mouse hepatocarcinogenesis by tetrachloroethylene via a metabolism-mediated pathway.

OTHER MODES OF ACTION

The operation of additional, non-PPARα-mediated mechanisms does not seem necessary to explain hepatocarcinogenesis by tetrachloroethylene but from a scientific point of view cannot be excluded. The question is whether evidence exists which supports a significant contribution of other MOAs to hepatocarcinogenesis.

Cytotoxicity

Tetrachloroethylene causes some hepatotoxicity in mice. It may be due to formation of reactive metabolites, including trichloroacetyl chloride, which have shown protein binding in rodents (Pähler et al. 1999; Green et al. 2001). However, hepatotoxicity has been found to disappear almost completely within 30 days (Philip et al. 2007), and the available long-term carcinogenicity studies revealed little evidence of hepatic damage or inflammation (NTP 1986; JISA 1993). Nevertheless, because the relation between cytotoxicity, inflammation, and cancer is not sufficiently understood, this point should receive attention in future studies. TCA also exerts little hepatotoxicity (Bull et al. 1990; DeAngelo et al. 1989). Overall, current evidence does not indicate that hepatotoxicity of tetrachloroethylene or TCA contributes to hepatocarcinogenesis to a substantial extent. Protein binding in humans was below the level of detection (Pähler et al. 1999).

Genotoxicity

Hypothetically, genotoxic activity could produce initiated hepatocytes, whose development to tumors might be promoted by TCA. Genotoxic activity could thereby enhance the carcinogenic potential of TCA. However, although some metabolites of tetrachloroethylene are genotoxic, there is no convincing evidence of genotoxic or mutagenic effects of tetrachloroethylene in vivo, and no initiating potential has been detected in appropriate assays (committee report, Chapter 5). Thus, a contribution of genotoxicity to hepatic-tumor formation by tetrachloroethylene is not supported by current evidence.

DCA as the Active Metabolite

As described in section on Relevance of TCA vs DCA, substantial contribution to PPARα-mediated tumor formation is unlikely. The potential MOAs of DCA include genotoxicity, but this activity is weak and probably not relevant at the low levels formed (IARC 2004).

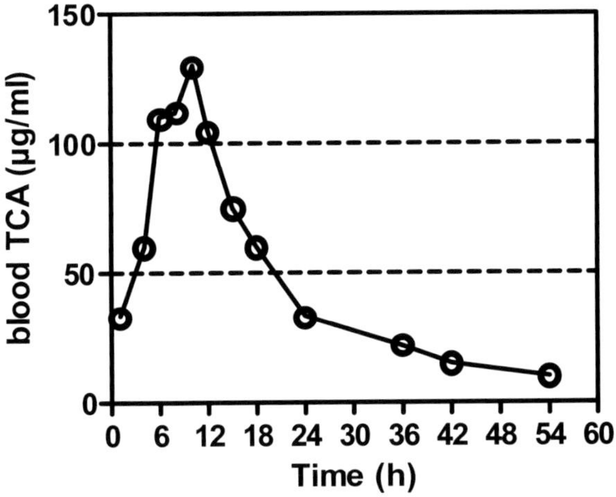

FIGURE 2A TCA concentrations in blood after single exposure of mice to tetrachloroethylene at 400 ppm for 6 hours. Source: Odum et al. 1988. Reprinted with permission; copyright 1988, *Toxicology and Applied Pharmacology*.

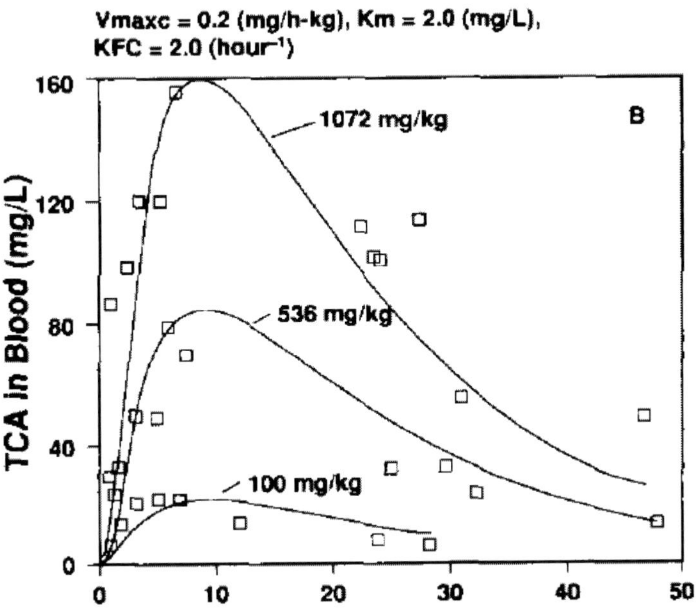

FIGURE 2B TCA concentrations in blood of male mice after single dose of tetrachloroethylene at 0.1, 0.536, and 1.072 mg/kg in corn oil by gavage. Experimental data shown as symbols; computer simulations shown as solid lines. Source: Gearhart et al. 1993. Reprinted with permission; copyright 1993, *Toxicology Letters*.

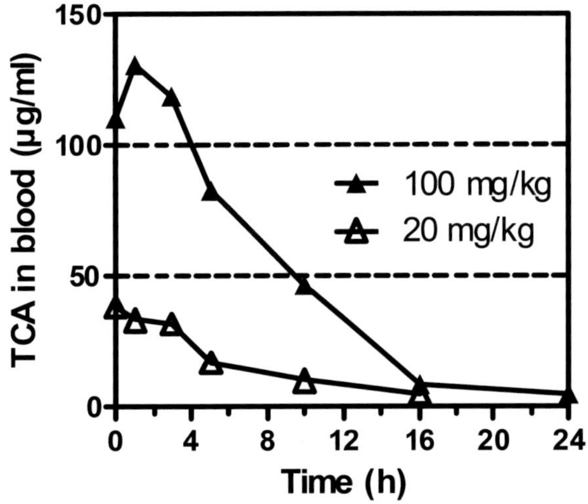

FIGURE 2C TCA concentrations in blood of male mice after single doses of TCA at 20 or 100 mg/kg by gavage. Data read from figure. Source: Larson and Bull 1992. Reprinted with permission; copyright 1992, *Toxicology and Applied Pharmacology*.

Other Mechanisms

Several other effects of tetrachloroethylene or TCA have been discussed as potential MOAs. Among these, changes in DNA methylation occur during PPARα activation. Therefore, that effect, although not specific for the PPARα MOA, does not necessarily support a contribution of other MOAs to hepatocarcinogenicity of tetrachloroethylene. TCA slightly transactivated mPPARγ, but this effect was much weaker than seen with PPARα and therefore is considered to have little or no relevance in mouse hepatocarcinogenesis. Importantly,TCA had no effect on human PPARγ (see section on PPARα Transactivation). In conclusion, there is no evidence available tosuggest that MOAs other than PPARα activation have asignificant impact on mouse hepatocarcinoma formation by tetrachloroethylene. Therefore, the weight of evidence supports the PPARα MOA.

SOME RECENT FINDINGS CONCERNING THE ROLE OF PPARα ACTIVATION IN MOUSE AND HUMAN HEPATOCARCINOGENESIS

The evidence suggesting that PPARα activation plays a causal role in rodent hepatic-tumor formation by many peroxisome proliferators but is not relevant for human hepatocarcinogenesis has been compiled in recent reviews (Klaunig et al. 2003; Meek et al. 2003; Peters et al. 2005; EU 2008; Corton 2008).

The authors of the IRIS draft present two recent publications that in their opinion raise questions about the causal relationship between activation of PPARα and rodent hepatic-tumor formation (p. 4-31). First, Yang et al. (2007) used transgenic mice (LAP-VP16PPARα) that target constitutively activated PPARα specifically at hepatocytes. The transgenic mice exhibited various PPARα-mediated effects—changes in fatty acid metabolism peroxisome proliferationand hepatocyte proliferation—but, surprisingly, not hepatic tumors after 1 year. Transgenic mice showed no hepatocyte hypertrophy and eosinophilia and no induction of proliferation of nonparenchymal liver cells. Those results indicate that PPARα-dependent induction of hepatocyte proliferation alone is not sufficient for hepatocarcinogenesis and that additional effects, such as activation of nonparenchymal cells, are required. Activation of Kupffer and other nonparenchymal cells had been found necessary for optimal induction of proliferation of normal and preneoplastic hepatocytes (Rose et al. 1997; Parzefall et al. 2001; Hasmall et al. 2001; Drucker et al. 2006). Thus, the study of Yang et al. does not refute the PPARα MOA but confirms and extends current knowledge.

Second, Ito et al. (2007) found that a low dose of DEHP (0.05% in diet) known to be noncarcinogenic in wild-type mice produced a low rate (26%) of hepatic adenomas in PPARα-null mice after 22 months. The tumors apparently were induced by oxidative stress and inflammation, as indicated by histopathologic changes and increases in 8-OHdG, NF-κB, and c-jun RNA, all of which were particularly high in the null mice. Activation of PPARα can have anti-inflammatory effects, resulting in higher vulnerability to tumorigenesis in PPARα-null mice (Ito et al. 2007). 8-OHdG was not increased after tetrachloroethylene or TCA (see earlier section on TCA as a Peroxisome Proliferator and Hepatocarcinogen in Mice). Thus, the results of Ito et al. suggest that DEHP, an agent unrelated to tetrachloroethylene, can induce (benign) hepatic tumors through a second, previously unsuspected PPARα-independent pathway. They do not contradict the causal role of PPARα activation in many instances of rodent hepatocarcinogenesis induced by peroxisome proliferators, which is supported by overwhelming evidence.

Some important new findings are missing in the IRIS draft. Thus, the generation of transgenic mice in which the mouse PPARα is replaced by the human counterpart provided substantial progress. The hPPARα mice were essentially resistant to hepatocarcinogenesis when fed a potent peroxisome proliferator (WY-14643) for 44 weeks, whereas corresponding wild-type mice developed tumors in 38 weeks. Gene-expression analysis for peroxisomal fatty-acid-metabolizing enzymes revealed that both receptors were functional. The findings suggest that structural differences between human and mouse PPARα are responsible for the different susceptibility of mice and humans to hepatocarcinogenesis by peroxisome proliferators (Morimura et al. 2006).

Furthermore, it was shown that induction of hepatocellular proliferation by peroxisome proliferators involves downregulation of the microRNA let-7c gene by mPPARα. That in turn allows increased expression of c-myc protein, which is essential for hepatocyte proliferation and tumor formation. Human PPARα apparently cannot suppress let 7c expression, and c-myc was not increased in hPPARalpha mice after WY-14643 treatment (Shah et al. 2007; Gonzalez and Shah 2008). Overall, the findings provide mechanism-based support for the concept that the PPARα MOA of rodent-hepatocarcinoma induction is not relevant to human hepatocarcinogenesis.

SUMMARY

This dissent has critically reviewed evidence related to MOAs of mouse hepatocarcinogenesis after exposure to tetrachloroethylene. The following conclusions can be drawn from findings in the literature:

1. TCA is the major metabolite in the body after exposure to tetrachloroethylene. DCA concentrations in blood and liver were lower than those of TCA by an order of magnitude, or DCA was completely undetectable.

2. TCA transactivates PPARα, while tetrachloroethylene does not. DCA also activates PPARα, but, because of its low occurrence, arguments related to the PPARα MOA should focus on TCA as the dominant active metabolite.

3. Effects of tetrachloroethylene and TCA associated with peroxisome proliferation were compiled and evaluated for consistency with the PPARα MOA as suggested by Klaunig et al. (2003). TCA induces the three key causal events, as well as peroxisome proliferation, and other associatedkey events. Data were generated in several studies, and dose-response and temporal relationships are consistent with the observation of tumors. The weight of evidence of this MOA was considered *strong* for TCA (in agreement with the National Research Council trichloroethylene committee) and *probable* for tetrachloroethylene although studies of PPARα-null mice are not available. Major support of the PPARα MOA of tetrachloroethylene rests on the role of TCA as the active metabolite.

4. Rats are less sensitive than mice to PPARα-mediated effects of tetrachloroethylene and do not develop hepatocarcinoma in response to tetrachloroethylene or TCA. That species difference can be explained by kinetic differences in TCA formation and availability in the target organ. In mice, formation of TCA is much higher and binding to plasma proteins much lower than in rats. Therefore, the mouse-rat difference can be explained by assuming that TCA is the active metabolite of tetrachloroethylene.

5. A key question is whether sufficient TCA is produced from tetrachloroethylene to induce peroxisome proliferation and tumor formation in the liver. To address that question, analytic data on blood and liver concentrations of TCA were collected from the literature. The data revealed that peak and AUC levels of TCA in mouse blood after tetrachloroethylene were similar to or even higher than those after TCA when carcinogenic doses of the two agents were compared. That constitutes direct evidence that TCA can be generated from tetra-

chloroethylene and be present in blood and target organ in amounts sufficient to induce peroxisome proliferation and hepatocarcinogenesis.

6. Analytic data from all of five available studies consistently demonstrate that absorption of TCA after oral application is incomplete and decreases with increasing dose. Moreover, published modeling work based on some of those studies suggests that only 25-10% of oral TCA bioavailable. The analytic and modeling data are not compatible with the estimate in the IRIS draft that 95% of oral TCA is absorbed—an estimate apparently not founded on experimental data. Apparently, the internal TCA doses derived from that estimate are too high. Consequently, the tumor yields predicted for tetrachloroethylene-derived TCA would be too low. Indeed, modeling studies taking into account the limited bioavailability of TCA suggest that TCA generated from tetrachloroethylene is sufficient to explain the incidence of hepatic tumors.

In conclusion, the weight of evidence clearly favors a key role of PPARα activation by TCA in tetrachloroethylene-induced mouse hepatocarcinogenesis.

7. The available evidence does not support a substantial contribution of other MOAs to hepatocarcinogenesis by tetrachloroethylene.

8. Transgenic mice carrying the human PPARα gene were found to be essentially resistant to hepatocarcinogenesis by a model peroxisome proliferator. This and other recent molecular data provide mechanism-based support for the concept that the PPARα MOA lacks relevance to human hepatocarcinogenesis.

COMMITTEE REBUTTAL

The committee greatly appreciates the dissenting member's thoughtful and careful review of the scientific literature and presentation of the arguments with respect to the MOA of tetrachloroethylene in mouse hepatic tumors and its relevance to humans. As noted by the dissenter and in Chapter 6 of the committee's report, the committee agrees that the EPA MOA characterization for hepatic cancer is inadequate and should be revised to provide a more focused and integrated analysis of the available evidence on tetrachloroethylene and its metabolites. The dissenter's statement is an attempt to provide an example of how such an analysis might be performed. The committee supports much of the dissenter's approach, but the dissenting member's conclusions go beyond those drawn by the full committee.

The dissenting member holds the opinion that PPARα mediation of tetrachloroethylene-induced hepatocarcinogenesis in mice is the plausible predominant MOA and that this MOA lacks relevance to human hepatocarcinogenesis. The committee believes that the arguments presented are reasonable and advises EPA to review the considerations presented by the member and the recent literature cited carefully. However, the committee does not support the apparent conclusions regarding mouse hepatic cancer that TCA is the sole carcinogenic metabolite of tetrachloroethylene, that the only MOA of TCA is peroxisome proliferation, and that there is unmistakable concordance in the carcinogenic potency of tetrachloroethylene in the National Toxicology Program and Japan Industrial Safety Association bioassays and the corresponding studies of TCA. Overall, the committee judges that many gaps in knowledge remain with regard to the MOA of tetrachloroethylene and that the relevance of the peroxisome-proliferator MOA to tetrachloroethylene-induced mouse hepatic cancer and to tetrachloroethylene-induced human hepatic cancer remains hypothetical and requires further rigorous testing.

The committee generally supports the comprehensive literature review and analyses conducted by the dissenting member and recommends that EPA use them when reassessing its own evaluation. However, there are aspects of the dissenter's analysis that the committee believes require more rigorous assessments before definitive conclusions can be drawn. They include the following:

- The committee does not agree that a role of DCA in tetrachloroethylene-induced hepatic carcinogenesis in mice can be ruled out solely on the grounds that it is detected at much lower concentrations than TCA in the blood and liver. First, there are few data on DCA formation from tetrachloroethylene. Second, there is some evidence that DCA is formed via a metabolic pathway that does not involve the liver. Third, there is some debate on whether DCA is formed from TCA. In Chapter 6, the committee stated that the conclusions regarding potential relevance or lack of relevance of DCA to hepatic carcinogenesis by tetrachloroethylene would be strengthened by the comparison of tetrachloroethylene hepatocellular-tumor data with predictions based on DCA carcinogenesis studies (in a way similar to that presented in Appendix 4A of the draft IRIS assessment). Such an analysis would provide a strong quantitative rationale for DCA's potential involvement, or lack thereof, in hepatic cancer.
- A more critical look at the quantitative differences in metabolic activation of tetrachloroethylene to TCA between mouse and rat, species that are generally believed to be almost equally sensitive to peroxisome proliferation, and in induction of hepatic cancer by other compounds in this class should be conducted by EPA. Chapter 6 recommends that EPA consider performing additional analyses with the rat data similar to those done with the mouse in Appendix 4A of the draft and including a table that shows the quantitative differences in affinity to mouse, rat, and human PPARα of both tetrachloroethylene and its key metabolites in comparison with the known peroxisome proliferators. Such analyses and data would greatly facilitate the discussion of quantitative differences between compounds and species.
- The committee supports the use of the weight-of-evidence analysis and the need for evaluation of the key events in hepatic carcinogenesis by tetrachloroethylene and its key metabolites. However, important knowledge gaps remain to be addressed with regard to key events in the PPARα MOA, especially those with causal and associative relationship to tumor formation and tetrachloroethylene or its key metabolites (see dissenter's statement and Chapter 6). Indeed, the committee is not yet convinced of the proof of the hypothesis that

the PPARα MOA is the sole MOA of tetrachloroethylene in inducing mouse hepatic cancer. Hence, it is premature to draw conclusions on the relevance of the PPARα MOA to tetrachloroethylene-induced human hepatic carcinogenesis.

- The committee agrees that the issues of TCA bioavailability, absorption, and blood and liver concentrations in various exposure scenarios are critical for the consideration of MOA of tetrachloroethylene. The current analysis by EPA is important but is inadequate in its current form. The committee recommends that EPA reconsider the analyses performed and consider using the data suggested by the dissenting member.
- The committee disagrees with the dissenter that the available evidence is sufficient to conclude that other MOAs are unlikely to contribute substantially to hepatocarcinogenesis by tetrachloroethylene. As noted in Chapter 6, the committee recommends that EPA strengthen and clarify the description of the degree, rather than the "significance," of the contribution of other plausible molecular events, in addition to activation of PPARα, to mouse hepatic tumors produced by tetrachloroethylene.
- The committee agrees with the dissenter that recent findings reported with PPARα-null mice (Ito et al. 2007; Takashima et al. 2008; Eveillard et al. 2009), PPARα humanized transgenic mice (Morimura et al. 2006), and hepatocyte-specific constitutively activated PPARα transgenic mice (Yang et al. 2007) are valuable contributions to the discussion of the relevance of the PPARα MOA in human hepatic carcinogenesis. The dissenter cites those studies to draw a conclusion that the PPARα MOA lacks relevance to human hepatocarcinogenesis. However, alternative conclusions that can be drawn from the studies mentioned above are that the short-term carcinogenesis studies in the PPARα-null mouse model have important limitations, that activation of PPARα is necessary but not sufficient for the development of mouse hepatic tumors, and that additional molecular events may be important parts of the peroxisome-proliferator MOA. Thus, the committee believes that it is premature to draw definitive conclusions regarding the relevance of the PPARα MOA to human hepatocarcinogenesis. In Chapter 6, the committee has encouraged EPA to strengthen the discussion of this matter in the draft IRIS assessment.

REFERENCES

Austin, E.W., J.M. Perrish, D.H. Kinder, and R.J. Bull. 1996. Lipid peroxidation and formation of 8-hydroxydeoxyguanosine from acute doses of halogenated acetic acids. Fundam. Appl. Toxicol. 31(1):77-82.

Benane, S.G., C.F. Blackman, and D.E. House. 1996. Effect of perchloroethylene and its metabolites on intercellular communication in clone 9 rat liver cells. J. Toxicol. Environ. Health 48(5):427-437.

Buben, J.A., and E.J. O'Flaherty. 1985. Delineation of the role of metabolism in the hepatotoxicity of trichloroethylene and perchloroethylene: A dose-effect study. Toxicol. Appl. Pharmacol. 78(1):105-122.

Bull, R.J. 2000. Mode of action of liver tumor induction by trichloroethylene and its metabolites, trichloroacetate and dichloroacetate. Environ. Health Perspect. 108(Suppl 2):241-259.

Bull, R.J., I.M. Sanchez, M.A. Nelson, J.L. Larson, and A.J. Lansing. 1990. Liver tumor induction in B6C3F1 mice by dichloroacetate and trichloroacetate. Toxicology 63(3):341-359.

Bull, R.J., G.A. Orner, R.S. Cheng, L. Stillwell, A.J. Stauber, L.B. Sasser, M.K. Lingohr, and B.D. Thrall. 2002. Contribution of dichloroacetate and trichloroacetate to liver tumor induction in mice by trichloroethylene. Toxicol. Appl. Pharmacol. 182(1):55-65.

Bull, R.J., L.B. Sasser, and X.C. Lei. 2004. Interactions in the tumor-promoting activity of carbon tetrachloride, trichloroacetate, and dichloroacetate in the liver of male B6C3F1 mice. Toxicology 199(2-3):169-183.

Corton, J.C. 2008. Evaluation of the role of peroxisome proliferator-activated receptor α (PPARα) in mouse liver tumor induction by tetrachloroethylene. Crit. Rev. Toxicol. 38(10):857-875.

DeAngelo, A.B., F.B. Daniel, L. McMillan, P. Wernsing, and R.E. Savage, Jr. 1989. Species and strain sensitivity to the induction of peroxisome proliferation by chloroacetic acids. Toxicol. Appl. Pharmacol. 101(2):285-298.

DeAngelo, A.B., F.B. Daniel, B.M. Most, and G.R. Olson. 1997. Failure of monochloroacetic acid and trichloroacetic acid administered in the drinking water to produce liver cancer in male F344/N rats. J. Toxicol. Environ. Health 52(5):425-445.

DeAngelo, A.B., F.B. Daniel, D.M. Wong, and M.H. George. 2008. The induction of hepatocellular neoplasia by trichloroacetic acid administered in the drinking water of the male B6C3F1 mouse. J. Toxicol. Environ. Health A 71(16):1056-1068.

Dees, C., and C. Travis. 1994. Trichloroacetate stimulation of liver DNA synthesis in male and female mice. Toxicol. Lett. 70(3):343-355.

Drucker, C., W. Parzefall, O. Teufelhofer, M. Grusch, A. Ellinger, R. Schulte-Hermann, and B. Grasl-Kraupp. 2006. Non-parenchymal liver cells support the growth advantage in the first stages of hepatocarcinogenesis. Carcinogenesis 27(1):152-161.

EPA (U.S. Environmental Protection Agency). 2008. Toxicological Review of Tetrachloroethylene (Perchloroethylene) (CAS No. 127-18-4) in Support of Summary Information on the Integrated Risk Information System (IRIS). External Review Draft. U.S. Environmental Protection Agency, Washington, DC.

EU (European Union). 2008. Risk Assessment Report on Tetrachloroethylene (PERC). CAS No. 127-18-4. EINECS No. 204-825-9. Draft Human Health Report. R021_0712_hh_SCHER. January 2008 [online]. Available: http://ecb.jrc.ec.europa.eu/documents/Existing-Chemicals/RISK_ASSESSMENT/DRAFT/R021_0712_env_hh.pdf [accessed Oct. 21, 2009].

Eveillard, A., L. Mselli-Lakhal, A. MOgha, F. Lasserre, A. Polizzi, J.M. Pascussi, H. Guillou, P.G. Martin, and T. Pineau. 2009. Di-(2-ethylhexyl)-phthalate (DEHP) activates the constitutive androstane receptor (CAR): a novel signalling pathway sensitive to phthalates. Biochem. Pharmacol. 77(11):1735-1746.

Ferreira-Gonzalez, A., A.B. DeAngelo, S. Nasim, and C.T. Garrett. 1995. Ras oncogene activation during hepatocarcinogenesis in B6C3F1 male mice by dichloroacetic and trichloroacetic acids. Carcinogenesis 16(3):495-500.

Gearhart, J.M., D.A. Mahle, R.J. Greene, C.S. Seckel, C.D. Flemming, J.W. Fisher, and H.J. Clewell, III. 1993. Variability of physiologically based pharmacokinetic (PBPK) model parameters and their effects on PBPK model predictions in a risk assessment for perchloroethylene (PCE). Toxicol. Lett. 68(1-2):131-144.

Goldsworthy, T.L., and J.A. Popp. 1987. Chlorinated hydrocarbon-induced peroxisomal enzyme activity in relation to species and organ carcinogenicity. Toxicol. Appl. Pharmacol. 88(2):225-233.

Gonzalez, F.J., and Y.M. Shah. 2008. PPARalpha: Mechanism of species differences and hepatocarcinogenesis of peroxisome proliferators. Toxicology 246(1):2-8.

Gonzalez-Leon, A., J.L. Merdink, R.J. Bull, and I.R. Schultz. 1999. Effect of pre-treatment with dichloroacetic or trichloroacetic acid in drinking water on the pharmacokinetics of a subsequent challenge dose in B6C3F1 mice. Chem. Biol. Interact. 123(3):239-253.

Green, S.M., M.F. Khan, B.S. Kaphalia, and G.A. Ansari. 2001. Immunohistochemical localization of trichloroacylated protein adducts in tetrachloroethene-treated mice. J. Toxicol. Environ. Health A 63(2):145-157.

Green, T. 2003. The Concentrations of Trichloroacetic Acid in Blood Following Administration of Trichloroacetic Acid in Drinking Water. Central Toxicology Laboratory, Alderley Park Macclesfield, Cheshire, UK (as cited in Sweeney et al. 2009).

Hasmall, S., N. James, K. Hedley, K. Olsen, and S. Roberts. 2001. Mouse hepatocyte response to peroxisome proliferators: Dependency on hepatic nonparenchymal cells and peroxisome proliferator activated receptor alpha (PPARalpha). Arch. Toxicol. 75(6):357-361.

Herren-Freund, S.L., M.A. Pereira, M.D. Khoury, and G. Olson. 1987. The carcinogenicity of trichloroethylene and its metabolites, trichloroacetic acid and dichloroacetic acid, in mouse liver. Toxicol. Appl. Pharmacol. 90(2):183-189.

IARC (International Agency for Research on Cancer). 2004. Some Drinking-Water Disinfectants and Contaminants, Including Arsenic. IARC Monographs on the Evaluation of Carcinogenic Risks to Humans, Vol. 84. Lyon: IARC.

Ito, Y., O. Yamanoshita, N. Asaeda, Y. Tagawa, C.H. Lee, T. Aoyama, G. Ichihara, K. Furuhashi, M. Kamijima, F.J. Gonzalez, and T. Nakajima. 2007. Di(2-ethylhexyl) phthalate induces hepatic tumorigenesis through a peroxisome proliferator-activated receptor alpha-independent pathway. J. Occup. Health 49(3):172-182.

JISA (Japanese Industrial Safety Association). 1993. Carcinogenicity Study of Tetrachloroethylene by Inhalation in Rats and Mice. Data No. 3-1. Kanagawa, Japan: Japanese Industrial Safety Association.

Klaunig, J.E., M.A. Babich, K.P. Baetcke, J.C. Cook, J.C. Corton, R.M. David, J.G. DeLuca, D.Y. Lai, R.H. McKee, J.M. Peters, R.A. Roberts, and P.A. Fenner-Crisp. 2003. PPARalpha agonist-induced rodent tumors: Modes of action and human relevance. Crit. Rev. Toxicol. 33(6):655-780.

Larson, J.L., and R.J. Bull. 1992. Metabolism and lipoperoxidative activity of trichloroacetate and dichloroacetate in rats and mice. Toxicol. Appl. Pharmacol. 115(2):268-277.

Laughter, A.R., C.S. Dunn, C.L. Swanson, P. Howroyd, R.C. Cattley, and J.C. Corton. 2004. Role of the peroxisome proliferator-activated receptor alpha (PPARalpha) in responses to trichloroethylene and metabolites, trichloroacetate and dichloroacetate in mouse liver. Toxicology 203(1-3):83-98.

Lumpkin, M.H., J.V. Bruckner, J.L. Campbell, C.E. Dallas, C.A. White, and J.W. Fisher. 2003. Plasma binding of trichloroacetic acid in mice, rats, and humans under cancer bioassay and environmental exposure conditions. Drug Metab. Dispos. 31(10): 1203-1207.

Mahle, D.A., R.J. Godfrey, G.W. Buttler, L. Narayanan, J.W. Fischer, and A. Taylor. 2001. Pharmacokinetics and Metabolism of Dichloroacetic Acid and Trichloroacetic Acid Administered in Drinking Water in Rats and Mice. AFRL-HE-WP-TR-

2001-0059. United States Air Force Research Laboratory, Wright Patterson Air Force Base, OH.

Maloney, E.K., and D.J. Waxman. 1999. trans-Activation of PPARalpha and PPARgamma by structurally diverse environmental chemicals. Toxicol. Appl. Pharmacol. 161(2):209-218.

Maronpot, R.R., T. Fox, D.E. Malarkey, and T.L. Goldsworthy. 1995. Mutations in the ras proto-oncogene: Clues to etiology and molecular pathogenesis of mouse liver tumors. Toxicology 101(3):125-156.

Meek, M.E., J.R. Bucher, S.M. Cohen, V. Dellarco, R.N. Hill, L.D. Lehman-McKeeman, D.G. Longfellow, T. Pastoor, J. Seed, and D.E. Patton. 2003. A framework for human relevance analysis of information on carcinogenic modes of action. Crit. Rev. Toxicol. 33(6):591-653.

Morimura, K., C. Cheung, J.M. Ward, J.K. Reddy, and F.J. Gonzalez. 2006. Differential susceptibility of mice humanized for peroxisome proliferator-activated receptor alpha to Wy-14,643-induced liver tumorigenesis. Carcinogenesis 27(5):1074-1080.

NCI (National Cancer Institute). 1977. Bioassay of Tetrachloroethylene for Possible Carcinogenicity. NCI-CG-TR-13. DHEW(NIH)77-813. U.S. Department of Health, Education, and Welfare, Public Health Service, National Institute of Health, Bethesda, MD [online]. Available: http://ntp.niehs.nih.gov/ntp/htdocs/LT_rpts/tr013.pdf [accessed Oct. 14, 2009].

NRC (National Research Council). 2006. Assessing the Human Health Risks of Trichloroethylene. Washington, DC: The National Academies Press.

NTP (National Toxicology Program). 1986. Toxicology and Carcinogenesis Studies of Tetrachloroethylene (Perchloroethylene) CAS No. 127-18-4) in F344/N Rats and B6C3F$_1$ Mice (Inhalation Studies). NTP TR 311. NIH Publication No. 86-2567. U.S. Department of Health and Human Services, Public Health Service, National Institutes of Health, Research Triangle Park, NC.

Odum, J., T., Green, J.R. Foster, and P.M. Hext. 1988. The role of trichloroacetic acid and peroxisome proliferation in the differences in carcinogenicity of perchloroethylene in the mouse and rat. Toxicol. Appl. Pharmacol. 92(1):103-112.

Pahler, A., J. Parker, and W. Dekant. 1999. Dose-dependent protein adduct formation in kidney, liver, and blood of rats and in human blood after perchloroethene inhalation. Toxicol. Sci. 48(1):5-13.

Parrish, J.M., E.W. Austin, D.K. Stevens, D.H. Kinder, and R.J. Bull. 1996. Haloacetate-induced oxidative damage to DNA in the liver of male B6C3F1 mice. Toxicology 100(1-3):103-111.

Parzefall, W., W. Berger, E. Kainzbauer, O. Teufelhofer, R. Schulte-Hermann, and R.G. Thurman. 2001. Peroxisome proliferators do not increase DNA synthesis in purified rat hepatocytes. Carcinogenesis 22(3):519-523.

Pereira, M.A. 1996. Carcinogenic activity of dichloroacetic acid and trichloroacetic acid in the liver of female B6C3F1 mice. Fundam. Appl. Toxicol. 31(2):192-199.

Pereira, M.A., and J.B. Phelps. 1996. Promotion by dichloroacetic acid and trichloroacetic acid of N-methyl-N-nitrosourea-initiated cancer in the liver of female B6C3F1 mice. Cancer Lett. 102(1-2):133-141.

Peters, J.M., C. Cheung, and F.J. Gonzalez. 2005. Peroxisome proliferator-activated receptor-alpha and liver cancer: Where do we stand? J. Mol. Med. 83(10):774-785.

Philip, B.K., M.M. Mumtaz, J.R. Latendresse, and H.M. Mehendale. 2007. Impact of repeated exposure on toxicity of perchloroethylene in Swiss Webster mice. Toxicology 232(1-2):1-14.

Reitz, R.H., M.L. Gargas, A.L. Mendrala, and A.M. Schumann. 1996. In vivo and in vitro studies of perchloroethylene metabolism for physiologically based pharmacokinetic modeling in rats, mice, and humans. Toxicol. Appl. Pharmacol. 136(2):289-306.

Rose, M.L., D.R. Germolec, R. Schoonhoven, and R.G. Thurman. 1997. Kupffer cells are causally responsible for the mitogenic effect of peroxisome proliferators. Carcinogenesis 18(8):1453-1456.

Schumann, A.M., J.F. Quast, and P.G. Watanabe. 1980. The pharmacokinetics and macromolecular interactions of perchloroethylene in mice and rats as related to oncogenicity. Toxicol. Appl. Pharmacol. 55(2):207-219.

Shah, Y.M., K. Morimura, Q. Yang, T. Tanabe, M. Takagi, and F.J. Gonzalez. 2007. Peroxisome proliferator-activated receptor alpha regulates a microRNA-mediated signaling cascade responsible for hepatocellular proliferation. Mol. Cell Biol. 27(12):4238-4247.

Stauber, A.J., and R.J. Bull. 1997. Differences in phenotype and cell replicative behavior of hepatic tumors induced by dichloroacetate (DCA) and trichloroacetate (TCA). Toxicol. Appl. Pharmacol. 144(2):235-246.

Stauber, A.J., R.J. Bull, and B.D. Thrall. 1998. Dichloroacetate and trichloroacetate promote clonal expansion of anchorage-independent hepatocytes in vivo and in vitro. Toxicol. Appl. Pharmacol. 150(2):287-294.

Sweeney, L.M., C.R. Kirman, M.L. Gargas, and P.H. Dugard. 2009. Contribution of trichloroacetic acid to liver tumors observed in perchloroethylene (perc)-exposed mice. Toxicology 260(1-3):77-83.

Takashima, K., Y. Ito, F.J. Gonzalez, and T. Nakajima. 2008. Different mechanisms of DEHP-induced hepatocellular adenoma tumorigenesis in wild-type and Ppar alpha-null mice. J. Occup. Health 50(2):169-180.

Templin, M.V., J.C. Parker, and R.J. Bull. 1993. Relative formation of dichloroacetate and trichloroacetate from trichloroethylene in male B6C3F1 mice. Toxicol. Appl. Pharmacol. 123(1):1-8.

Toraason, M., J. Clark, D. Dankovic, P. Mathias, S. Skaggs, C. Walker, and D. Werren. 1999. Oxidative stress and DNA damage in Fischer rats following acute exposure to trichloroethylene or perchloroethylene. Toxicology 138(1):43-53.

Walgren, J.E., D.T. Kurtz, and J.M. McMillan. 2000. Expression of PPAR(alpha) in human hepatocytes and activation by trichloroacetate and dichloroacetate. Res. Commun. Mol. Pathol. Pharmacol. 108(1-2):116-132.

Yang, Q., S. Ito, and F.J. Gonzalez. 2007. Hepatocyte-restricted constitutive activation of PPAR alpha induces hepatoproliferation but not hepatocarcinogenesis. Carcinogenesis 28(6):1171-1177.

Zhou, Y.C., and D.J. Waxman. 1998. Activation of peroxisome proliferator-activated receptors by chlorinated hydrocarbons and endogenous steroids. Environ. Health Perspect. 106 (Suppl. 4):983-988.